ATOMIC PHYSICS

ATOMIC PHYSICS

Dr. A.K. Singh

RANDOM PUBLICATIONS

NEW DELHI - 110 002 (INDIA)

Atomic Physics

ISBN 978-93-51117-44-5

Published in 2015 in India by

RANDOM PUBLICATIONS

4376-A/4B, Gali Murari Lal, Ansari Road
New Delhi-110 002
Phone: +9111-43580356, 23289044
E-mail: randomexports@gmail.com; sales@randompublications.com;
info@randompublications.com

Reprinted 2026

Type Setting by: Friends Media, Delhi-110089

Preface

The atom is the smallest unit that defines the chemical elements and their isotopes. Most matter on Earth is made up of atoms. The size of atoms is measured in picometers (trillionths of a metre). A single strand of human hair is about one million carbon atoms wide. Every atom is composed of a nucleus made of protons and neutrons (hydrogen-1 has no neutrons). The nucleus is surrounded by a cloud of electrons. The electrons in an atom are bound to the atom by the electromagnetic force, and the protons and neutrons in the nucleus are bound to each other by the nuclear force. Over 99% of the atom's mass is in the nucleus. The protons have a positive electric charge, the electrons have a negative electric charge, and the neutrons have no electric charge. Normally, an atom's electrons balance out the positive charge of its protons to make it electrically neutral. If an atom has a surplus or deficit of electrons, then it will have an overall charge, and is called an ion. The number of protons in the nucleus determines what chemical element the atom belongs to (e.g. all copper atoms contain 29 protons). The number of neutrons determines what isotope of the element it is. The electron cloud of the atom determines the atom's chemical properties and strongly influences its magnetic properties.

Atoms can attach themselves to each other by chemical bonds to form molecules, network solids, metal alloys, crystals, and other solid solutions. The tendency for atoms to bond and break apart is responsible for most of the physical changes we observe in nature, and this is studied by the science of chemistry. Atoms and sub-atomic particles behave in peculiar ways that cannot be explained through the classical laws of physics. The field of quantum mechanics was developed to explain the structure and behaviour of atoms. Not all matter is made up of atoms, but atoms do comprise all the types of matter than can be seen and touched. Astronomical observations indicate that most of the Universe's matter is "dark matter", composed of particles of a currently unknown type. Atomic physics is the field of physics that studies atoms as an isolated system of electrons and an atomic nucleus. It is primarily concerned with the arrangement of electrons around the nucleus and the processes by which these arrangements change. This includes ions as well as neutral atoms and, unless otherwise stated, for the purposes of this discussion it should be assumed that the term atom includes ions. The term atomic physics is often associated with nuclear power and nuclear bombs, due to the synonymous use of atomic and nuclear in standard English.

The book guides the reader to the research frontiers in atomic physics with heavy ions, giving a concise image of the advanced experimental and theoretical methods exemplified by the most recent results. This useful book presents a comprehensive overview of the spectacular advances seen in atomic physics during the last 50 years.

I thank all members of my team who have helped in the preparation of the book. My special thanks go to "Random Publications" who have published the book.

— Dr. A.K. Singh

Contents

1. Introduction 1

Atomic Modelling 1

Liquids 11

Liquids: Fluid Pressure/Surface Tension 13

Complications: Viscous Flows and Compressible Flows 22

Speculative Atomism 25

2. Models of Atom 27

Atomic Spectra 27

The Balmer Series 30

Models of the Atom 31

The Bohr Atom 35

Reality of Atoms 44

Quantized Atom 46

Molecular Motions and Energy 47

Spin-spin Splitting—A Closer Look 53

Etranslation, Erotation, Evibration 56

Electron Motion in Molecules—Eelectrons 67

The Correspondence of Colour and Spectrum 71

Relationship of Structure to the Electronic
Absorption Spectrum 72

Atomic Structure and Bonding 82

The Laws of Falling Bodies 83

3. Evaluation of the Atomic Concept 92

The Limits of the Divisibility of Matter 98

Contributions of Atomic Physics 106

Fundamental Forces 109

Symmetries and Particle Physics 111

Importance of Symmetries: Noether's Theorem 112

Internal and Space-time Symmetries 113

Discrete Symmetries 113

Continuous Symmetries 116

4. Nuclear Properties 130

Radioactive Decay 130

Alpha Decay 141

Beta Decay 144

Gamma Ray 147

Double Beta Decay 154

Double Electron Capture 156

Internal Conversion 157

Isomcric Transition 160

Cluster Decay 160

5. Nuclear Energy 164

Dissipation 164

Photovoltaic Effect 165

Photoelectric Effect 165

Bioenergy 175

Fossil Fuel 177

Global Dimming 183

Nuclear Fuel 190

Binding Energy 202

6. Nuclear Forces 206

Nuclear Reaction 206

Nuclear Force 211

Strong Interaction 215

Natural Nuclear Fission Reactor 218

Big Bang 221

Cosmic Microwave Background Radiation 230

Features, Issues and Problems 232

The Future According to the Big Bang Theory 237

Speculative Physics Beyond Big Bang Theory 238

Nuclear Reactor Physics 239

Nuclear Fission 246

Fission Reactors 254

7. Atomic Structure of Elements 265

Atomic Nucleus 265

Background Radiation 270

Radon 274

Magnetic Resonance Imaging 276

MRI versus CT 287

Economics of MRI 287

Radio Frequency Energy 289

Three-Dimensional (3D) Image Reconstruction 295

Virtual Particle 296

Manifestations 298

Spontaneous Fission 302

Neutron Emission 303

Positron Emission 303

Proton Emission 304

Electron Capture 305

Neutron Capture 306

Bibliography 309

Index 311

Chapter 1

Introduction

Atomic Modelling

Matter and Atoms

The idea that matter could be broken down into submicroscopic individual units, known as "atoms", goes back to antiquity, though the concept didn't start to acquire scientific momentum until 1808, when the English scientist John Dalton proposed it to account for the results of his studies of chemical reactions. Today, it is known that there are about a hundred different types of atoms with different weights and other properties. These "elements" include hydrogen, the lightest and most common element; carbon, which makes up soot; silicon, one of the constituents of sand; metals like gold, iron, and lead; the gaseous elements oxygen and nitrogen; and many others.

Atoms could also be combined to form "molecules". For example, if the element oxygen is burned with the element hydrogen, each oxygen atom forms "chemical bonds" with two hydrogen atoms to form water molecules.

It was the distinct and predictable patterns of combinations of atoms into molecules that led Dalton to propose the atomic theory. Large numbers of atoms can combine into a single molecule, particularly in the case of the complex carbon-based "biomolecules" that make up living organisms, and the materials made from atoms and molecules can be arranged into large-scale systems, from stones to machines to flowers to human beings.

Detailed discussion of chemical reactions, in which atoms are converted into molecules or molecules are broken down into atoms, needs to be reserved for a document on chemistry. However, it is useful to comment here that such chemical reactions observe the fundamental law of energy conservation as obeyed in physics in general. Reactions may be "exothermic", releasing energy stored in chemical bonds, or "endothermic", absorbing energy stored into chemical bonds; but in any case the total energy before and after the reaction will remain the same. Only the forms are changed. For example, water can be broken down into hydrogen and oxygen by running an electric current through it; this is an endothermic process. A spark will then cause the hydrogen and oxygen to explosively recombine; this is an exothermic process. On a broader canvas, a tree grows by using sunlight and raw materials such as water and carbon dioxide to drive its complicated biological processes, which in sum amounts to an endothermic process. The wood can be burned to release its energy, which is an exothermic process.

Oddly, although atomic theory now seems to be an essential element of classical physics, it wasn't completely accepted by physicists until the early 1900s. Ironically, this was at the very time that the doctrines of classical physics would begin to be seriously challenged. Atoms had been originally thought of as indivisible, but even before the final triumph of atomic theory physicists were beginning to realise that atoms themselves were not indivisible and had an internal structure. They gradually realised that an atom had a heavy central "nucleus" that had positive electric charge, surrounded by light orbiting "electrons" that had a negative electric charge. Electrons could be stripped from atoms to be sent through wires as electrical currents.

If an atom lost its normal complement of electrons, it became a positively charged "ion". Atoms could also acquire an excess of electrons, becoming negatively charged ions. Such charge imbalances were one mechanism that contributed to the bonding of atoms into molecules, though subtler processes were involved as well. Physicists soon found that the interior of the atom was a complicated place where things happened that were impossible at a larger scale, and this simple model of a "classical atom" had to be revised. The revisions lead into the realm of modern physics and so are beyond the scope of this document. The classical atom still provides a useful basis for the discussion of the fundamental properties of matter.

Gross Properties of Solids

Solid matter covers a wide range of forms, everything from a compacted block of trash containing every form of household rubbish to a glittering diamond, consisting of carbon atoms arranged in an orderly, uniform "crystalline" arrangement. At a high level of abstraction, chunks of matter have certain properties that can be characterized and measured, for example colour; density, or ratio of mass to volume; and various measures of strength, or the ability of that chunk of matter to retain its structure or "integrity" when subjected to various forces. Obviously these things are much easier to characterize for, say, a diamond than for a block of compacted trash.Solid matter also has certain properties of "scale" that are independent of what the matter is made up of. For example, let's take a set of cubes of solid matter without regard to what the matter is. Suppose one cube is a centimeter on a side. Now let's suppose Dexter stacks up the cubes to make a bigger cube that is two centimeters on a side.

This doubles the linear dimensions of the cube. The surface area of the cube has expanded by a factor of four, while the volume and mass of the cube has expanded by a factor of eight. In more formal terms, doubling the linear dimensions of an object squares its surface area and cubes its volume and mass. 1950s horror movies such as Them envisioned monster ants, but in fact such scaling effects prevent insects from becoming very large. For one thing, they don't have lungs, and have to acquire oxygen through pores in their chitinous skins.

Doubling their size would increase their volume and mass twice as fast as it would increase their surface area, halving their ability to obtain oxygen for that mass, and at a certain size a giant insect would simply suffocate.A more important scaling issue that rules out giant insects is the issue of "compressive strength", or the ability of a structural support to bear weight placed on top of it.

The compressive strength of a structural support, such as a column that holds up a building, is proportional to the cross-sectional area of a horizontal slice through the column.A cross section is just one face of the surface of an object, and so like surface area increases with the square of a linear increase in size. However, the mass of the building increases with the cube of a linear increase in size. Double the linear dimensions of a building and the ability of a column to support the building's weight is increased by four, while the mass and the load on the column is increased by eight. The column now has to proportionately bear twice as much load as it did at the smaller scale. This means that more, or disproportionately large, columns must be used to support a larger building. Insects generally have spindly bodies and legs, while an elephant has great stumpy massive legs. If an insect was scaled up to large size, it would simply collapse of its own weight. However, horror movie fans can still take comfort that deadly venomous or parasitic insects are not ruled out by the laws of physics.

Strength of Materials

These considerations of scale lead back to the notion of strength of materials, which in turn leads to consideration of how materials are used for building structures. Although builders always had useful concepts of what could or could not be done when building structures, the first person to try to perform scientific experiments to determine the strength of materials was the Italian polymath Leonardo da Vinci.

The first formal text on the subject of materials was actually published late in life by Galileo in 1638, titled dialogues concerning two new sciences. He was under house arrest at the time, having antagonized the Vatican by promoting the notion that the Earth went around the Sun and not the reverse, and materials seemed like a noncontroversial subject not likely to get him into further trouble. Galileo's dialogues was not particularly correct in all details the basic laws of physics needed hadn't been invented just yet but the book did get people thinking. Isaac Newton had little interest in the subject of materials, but his rival Robert Hooke was fascinated by the subject, examining the ways in which materials responded to forces, in particular coming up with Hooke's law.

This was mentioned in a previous chapter as:

$$\text{Force} = \text{stiffness} \times \text{compression}$$

It was stated then as applying to a spring, but it also applies to structural element made of hard metals, stone, and glass. It does not in general apply to flexible materials like rubber or tendon, which stretch easily up to a limit but then yield diminishing returns. Hooke's work was the first to at least imply the notions of "stress" and "strain".

Stress is a force applied to a structural element. Such forces include:

- Compression forces, which squeeze the element in on itself.
- Tension forces, which try to pull the element apart.
- Shear forces, like we exert on a piece of cardboard to rip it in half.
- Torsion or twisting forces.
- Bending forces.

Stress is measured in terms of force per unit of area, or what is simply known as "pressure". The formal metric unit of pressure is the "pascal (pa)", which is a newton per square meter, named after the great French mathematician and physicist Blaise Pascal. Since a pascal is a somewhat small unit, it is often expressed in terms of "kilopascals (kpa)" thousands of pascals or "megapascals (Mpa)" megapascals. In compression and tension forces, the force is applied directly through a cross-section of a structural element, and so the stress is measured in terms of Mpa.

In shear force, the force is applied along a plane through a structural element, and so stress is measured in terms of Mpa over the area of the interface between the two sections of the structural element. Torsion is similar to shear but involves a torque over the area of the interface. Bending is complicated, since it involves a shear force down through the structural element, a tension force along its bottom, and a compression force along its top. In practice, structural elements may be subjected to a number of these stresses at the same time. A structural element under stress will exhibit a "strain", that is, a lengthening, given in a simple proportion. Up to a certain limit, once the stress is removed, the structural member will return to its original length. This is called "elastic" behaviour".

It was certainly implied by Hooke, but the idea was developed further by Thomas Young, who published a paper in 1807 in which he defined the concept of "elastic modulus" or "Young's modulus". In modern terms, this is the ratio of stress of a particular material to its strain:

$$\text{elastic modulus} = \text{stress/strain}$$

It is somewhat more intuitive to express this as:

$$\text{strain} = \text{stress/elastic modulus}$$

Since strain is a simple proportion, the units of elastic modulus are still Mpa. For materials that stretch easily. the elastic modulus has a low value; for example, rubber has an elastic modulus of 7. Given these definitions of stress and strain, then Hooke's law can be rephrased as follows:

$$\text{force} = \text{stress} \times \text{strain}$$

Hooke's law also leads to the concept of "strain energy", or the potential energy contained in a strained structural element, which is similar to the equation for energy in a stretched spring:

$$\text{strain energy} = \text{volume} \times (1/2) \times \text{stress} \times \text{strain}^2$$

Once the stress on a structural member exceeds a certain limit, the member will be permanently strained out of shape, and will break if the stress is increased. Not all materials respond to forces in such a nice graded way. They may not deform at all up to the level of force where they snap. Such materials are said to be "brittle". Glass is a good example, since any practical experience shows how difficult it is to bend a glass rod without breaking it. Materials are said to have a "breaking strength" or "tensile strength", rated in the tension stress that causes them to snap. The concept goes back at least to Galileo, who performed tests in breaking strengths, though he didn't have the theoretical tools needed to describe it well.

Typical tensile strengths include:

High strength steel	1,500 Mpa
Ordinary commercial steel	400 Mpa
Cast iron	~200 Mpa
Aluminum alloys	~300 Mpa
Hemp rope	80 Mpa
Bone	40 Mpa
Nylon fibre	1,000 Mpa
Kevlar fibre	2,700 Mpa

Not all materials respond to different forces in the same way. Although metals are strong in both compression and tension, stone is very strong in compression but is weak in tension. Wood can be split easily by shear forces applied along the line of its grain, but will resist larger shear forces applied at a right angle to its grain. Forces acting consistently on a structural element over a long period of time may cause a slow deformation, or "creep", that can lead to failure even if the elastic limit is never exceeded. Similarly, a cycle of changing stresses repeated over and over may cause "fatigue" that will eventually lead to the failure of a structural element.

Although Hooke got the ball rolling, the notions of stress and strain were not articulated in their modern form until 1822, in a document published by a by a French mathematician named Augustin Cauchy. It would take time for Cauchy's work to sink in, since he wrote in a theoretical style that was more or less foreign to engineers, but now his concepts are regarded as one of the major foundations of structural design.

The underlying idea of all structural design is balance of stresses. If the stresses imposed on a structural element exceed the strength of the element, it will collapse. The stresses are set up by what engineers refer to as "loads". Some of the loads on a structure are obvious: the "dead load" set up by the weight of the structure, and the "live load" set up by the people, furnishings, equipment, and materials contained in the structure. These are known as "static loads" because they don't change, or don't change much, over time.

Others loads are not quite so obvious. There are the "hidden" loads, with stresses set up by the expansion or contraction of structural elements due to changes in temperature, or by unequal settling of the building's foundations. Structures may also be subjected to rapidly changing "dynamic" loads, such as those caused by winds and earthquakes. Dynamic loads can be particularly treacherous since they are somewhat unpredictable. In one famous example, the designer of the Citicorp skyscraper in New York City received a letter from a student who was curious about the building's stability in winds. The architect replied that the building's design had been analysed to ensure that it would stand up to storm winds and that it was safe.

However, after sending his reply, the architect realised that the analysis had been performed with the assumption that the winds would be perpendicular to the faces of the building. He re-did his analysis to see what would happen if the winds were at an angle, and to his shock found out that a severe windstorm would generate torsion forces that would bring the skyscraper down in the middle of Manhattan. The Citicorp building's steel framework was hastily reinforced.

Design of Structures

Structures can be divided into two classes: compressive structures, in which the structural elements are stacked on top of each other; and tensile or "suspension" structures, in which the structural elements are suspended from supports. One of the simplest examples of a compression structure is a brick wall, with the bricks piled on top of each other. One of the simplest examples of a tensile structure is a circus tent, with the canvas "big top" held up by ropes hung from a central pole. Of course, in practice most structures feature both compressive and tensile elements. The first really big structures were compression structures, since it was difficult to build durable large structures of wood, and stone doesn't lend itself to tensile structures. The major problem with stone compressive structures is that stone is very heavy, or more specifically has a low "strength to weight" ratio.This means that the weight of such a structure grows very quickly as it the structure grows bigger, and walls have to be made thicker and thicker at the base so the structure doesn't collapse. This approach ultimately leads to the simplest of all large stone structures, the pyramid, built not only in Egypt but in China and in the Pre-

Columbian New World as well. Pyramids are an impressive but purely brute-force design. Builders were also learning more sophisticated tricks working with wood, the other classic building material. Wood was used to construct ordinary buildings, such as homes, and structures such as bridges. The earliest wooden bridges no doubt consisted just of two beams set up across a stream and covered with planks. The major problem with this is that such a bridge is limited in length. It is supported at the ends but not in the middle, and as it gets longer so does the "lever arm" of the half-spans of the bridge, meaning that a person stepping on the middle of the bridge places a proportionately greater stress on it. If the bridge was over the top of a shallow stream, this problem could be addressed by driving pilings in the middle of the stream and fastening the middle of the span to them, but if the stream wasn't shallow or if the bridge was over the top of deep canyon, such a fix wasn't possible.

There were two possible solutions. The first was the rope "suspension bridge", a tension-based design. Wooden or stone towers could be constructed on both sides of the canyon, with heavy ropes slung across their tops and anchored to the ground. A wooden walkway could then be suspended from the ropes. This type of bridge has been around for millennia, and is thoroughly familiar from B-fiction adventure movies since they are excellent sites for dramatic action scenes. They always seem to be in a frighteningly bad state of repair, and the waters beneath are often full of crocodiles, piranhas, or other predatory creatures. For present purposes, a second option is more interesting. Suppose Dexter has a wooden bridge and wants to brace up the middle of the span from each end. He can do this by fitting a beam from the bank to the midspan at an angle. This allows the forces directed downward at midspan to be shunted towards the banks. Since this sets up a sideways force on the two support beams that tends to splay them apart from the middle, it is also useful to link them together at the bottom with a third beam, forming a triangular support structure known as a "truss". The truss is a basic structural form, and is composed of various triangular structural subunits linked together.

The triangular shape is very useful as it will stand up to stresses imposed in different directions, and it is also employed in the "cross-bracing" of structures used to keep them from falling over in heavy winds and the like. The truss is of course in as wide use as it ever was; one of the tidiest examples is the metal high-voltage transmission line tower, which is a lightweight structure made up of little triangular truss elements. The problem with wooden bridges is that they didn't tend to last very long. Stone is far more durable, but it does not lend itself to construction of trusses. However, the same underlying principle of diverting stresses to the sides could be used on stone bridges as well, by building them with curved structures called "arches".

The arch did imply the provision of supports to deal with the sideways loads it set up. This was generally simple with bridges, as they might be built between stony riverbanks or, if such weren't available, heavy footings could be constructed on each bank. This could be a tricky matter for other structures. The Gothic cathedrals of Europe, with their high arched or "vaulted" ceilings, were built with "flying buttresses", structures attached to the outside of the building to provide the necessary lateral support. The "dome" was an extension of the arch, rotated around in a full circle.

Like the arch, the dome required lateral support to handle the horizontal stresses. Islamic architecture, which demonstrates a certain fondness for domes, often involves half-domes or "semi-domes" ringing the main dome to distribute sideways forces down to the heavy walls of the buildings. Although complete spheres are not ordinarily used in buildings, they are an extension of the notion of an arch or dome, in which compression is distributed around the sphere. An egg can be surprisingly hard to crack if subjected to balanced compression forces try carefully squeezing a hard-boiled egg and spheres are often used in the crew-carrying elements of deep-ocean submersibles, since they can withstand extreme pressures.

Most basic elements of classical building design, like the arch and dome, were in place by Roman times. Major advances in structural design had to wait until the development of large-scale metalworking in the 19th century. Iron and then steel structural elements, with their strength in both compression and tension and their high strength-to-weight ratio, made possible great steps forward in building design.

Classical construction forms, such as trusses and arches, could be built with such metal beams. The beams were generally in the form of "I-beams", which consist of a length of metal with a long flat horizontal plate on the bottom, a similar long flat horizontal plate on top, and a vertical plate connecting the two horizontal plates. A beam is subjected to compression on top and tension on the bottom, and the I-beam is designed to provide the maximum cross-sectional area on bottom and top to resist these stresses. Such techniques could be used to build new sorts of structures, such as the modern skyscraper. The poor strength-to-weight ratio of stone made very tall buildings impossible, but metal structural elements allowed buildings to be made much taller. A skyscraper consists of a "cage" of steel girders that supports the floors and the walls.

The spectacular explosive demolition of such structures, which usually makes local or even national TV news, is based on the sequential detonation of charges that cut critical structural supports of the cage in a particular order, causing the building to cave in on itself in a heap if the event has been properly planned. Metal building elements also allowed the development of large truss bridges, with a latticework of metal beams to allow distribution of stresses. It also led to the development of modern suspension bridges in which the roadbed is hung from huge cables slung over high towers and anchored to heavy concrete "anchors" at each end and "cable-stay" bridges in which the roadbed is held up by cables connected directly to the towers.

Concrete was around in Roman times, though the recipe was lost in the Middle Ages and not rediscovered until the 19th century. Modern building techniques make it far more useful than it was to the Romans. Using steel reinforcing rods, it is commonly used for foundations, walls, and other structural elements under compressive loads. However, "prestressed" concrete elements can also be used to handle tension loads that would cause ordinary reinforced concrete to crack. A prestressed concrete structural element is cast in a frame that supports steel cables under tension and laced through the concrete. When the concrete hardens or "cures", the frame is removed, and the tension of the cables within the concrete matrix prevents it from cracking under tension loads.

There is a balance in the design of structures in the number of structural elements required. Reducing the number of elements reduces complexity, generally reducing the labour of assembly and cost. However, reducing the number of elements also usually means larger elements, which may be difficult to handle, and also more prone to damage.

Suppose a dome is made up of curved panels; the fewer the number of panels, the larger the panels, the greater the load each panel has to carry, making each panel more prone to cracking and also making replacement of the panels more troublesome. Ofcourse, if there are a very large number of small panels, installing them will be laborious, and trying to make sure the dome doesn't leak in the rain through the seams between the panels may be difficult. There are cases where the benefits of "subdivision" are obvious. The cables used to hold up a suspension bridge are each made of large numbers of individual strands; the breakage of a few strands has little effect on the strength of the cable as a whole. It is also possible to construct structural elements as "laminates", using layers of materials that are bonded together with adhesives.

Plywood is a common example of a laminate, and panel structures may be made as "sandwiches", with a lightweight core and stiff surface paneling, providing an improved strength to weight ratio. Simple corrugated cardboard is a common example of such a sandwich material; much the same approach can be used with metal panels as well, with a corrugated or honeycomb fill. However, laminates do present the danger of becoming separated or "delaminated" under stress. Similar principles applies to the types of "composite" materials that consist of threads of one material embedded in a matrix of another material. Examples include glass or graphite fibres embedded in plastic, or "whiskers" of silicon carbide embedded in metal.

Any failure of a few threads does not do much to undermine the integrity of the structural element made of the composite, and the threads prevent the propagation of cracks. Research is now being conducted on also incorporating sensors in composites in the form of threads to give indication of when a structural element is about to fail, but such "smart" materials are not in common use yet.

Liquids

Liquids have some properties similar to those of solids, such as colour and density, but of course they differ in their ability to flow and to take the shape of any container they are poured into. They possess "viscosity", which is a measure of their ability to flow. A "thick" liquid that is hard to shove through a pipe has a high viscosity, while a "runny" liquid that flows quickly through a pipe has a low viscosity. It is obvious that it is much easier to get vinegar out of a squeeze bottle than honey. Viscosity is related to the concept of "fluid friction": concepts of friction and drag apply to liquids, and a liquid with a high viscosity generates a high level of fluid friction.

Another obvious characteristic of liquids is that a body of liquid will seek a common level. A pond undisturbed by wind may be "smooth as glass", as the saying goes, and it will be perfectly level, at least if it's small relative to the curvature of the Earth. It has to be noted here that in more general terms, the angle of the surface of a liquid will be at a right angle to the direction of the force applied to a liquid, so water in a spinning bucket will tend to "climb up the walls" of the bucket. An undisturbed pond will be flat because it is only subject to the pull of gravity, which is straight down. If pipes of various diameters are stuck through the surface of this flat pond, it only takes a moment's thought to realise that the water level is the same no matter the diameter of a pipe. Yet another obvious property of liquids is that not only can they flow through pipes, objects can flow through or over them. This leads to the concept of "buoyancy", the tendency of objects to float or sink in a liquid.

This is unsurprisingly due to the density of the object relative to the liquid. A block of wood is less dense than water and floats on water, while a rock is more dense than water and sinks. The sunken rock will displace a volume of water equal to the volume of the rock. The floating block of wood has a slightly subtler behaviour. It will sink into the water under the influence of gravity until it displaces a volume of water equal to its own weight. At that point, it will reach an equilibrium of forces and sink no further.

This is known as "Archimedes' Principle", since it was devised by the Hellenic scientist Archimedes (287:212 BC). The forces supporting this floating block of wood will focus upward and inward on it, summing to a point known as the "centre of buoyancy". The centre of buoyancy is always above the centre of mass of a floating object; the farther the distance between the centre of buoyancy and the centre of mass, the more stable the object that is, the less its tendency to capsize.

According to tradition, Archimedes discovered this fact while getting into a bathtub, and used it to determine for the authorities if a crown being sold by a metalsmith was actually made of gold or a debased metal. Archimedes weighed the crown, then determined its volume from the amount of water displaced, and from that determined its density, which could be compared with that of pure gold.

The story says the metalsmith was found guilty and executed, but the tale is probably a myth: it is a bit unlikely that Archimedes, using the instruments available to him, could have performed the required measurements with enough accuracy to have detected a fraud unless the debasement was so gross that it would have been obvious by simply inspecting the crown. One simple example of the practical application of Archimedes' principle is the "Plimsoll line", which is a marking painted on the hull of a vessel in the form of a circle or diamond cut by a horizontal line. If the ship is sitting so low that the Plimsoll line is underwater, the ship is overloaded.

Another interesting example is provided by the question: if all the ice at the North Pole melted, how much would the sea level rise? From a simple consideration of buoyancy, the answer is: not at all. The polar ice displaces a mass of water equivalent to itself, and so when it melts, there is no overall change in the water level. Of course, the ice melting off of Greenland and Antarctica would raise the sea level, and there's also the consideration of the extremely unpredictable ways that an ice-free Arctic Sea might affect global climate.

Incidentally, if a more dense liquid is poured into a less dense liquid, as long as the two liquids don't have a tendency to mix, the two will separate, with the less dense liquid on top and the more dense liquid on bottom. The classic example of this is oil and water. Considerations of buoyancy govern the operation of a submarine. A submarine cruising on the surface of the ocean has an average density somewhat less than that of water. To submerge, it has to take in water in "ballast tanks", increasing its density to more than that of water and causing it to sink. To rise again to the surface, it has to purge the ballast tanks by blowing air into them, forcing the water out and lowering the submarine's density.

Once underwater, a submarine needs to maintain a density comparable to that of water so that it will not tend to either rise or sink. This issue is somewhat complicated by the fact that the submarine's hull is compressed as the vessel dives deeper, reducing the volume of the hull and so increasing the submarine's density. In any case, once the submarine reaches the desired cruise depth, it can then maneuver freely by essentially "flying" underwater, using the "diving planes" or "foreplanes" that are mounted on nose or the "sail" (what's sometimes called the "conning tower") of the submarine, along with the "tailplanes", which are generally arranged in a vertical cross pattern.

However, there is an added complication for a submarine moving underwater. It is not firmly supported, and so if its centre of gravity shifts from the middle of the vessel, it will tend to tip down or up, making it difficult to control. This issue, known as "trim", was a particular problem for early submarines. It was solved by adding an additional set of "trim tanks", in addition to the ballast tanks. There is a trim tank in the forward section of the submarine and a trim tank in the rear section of the submarine, with trim maintained by adjusting the relative amounts of water in these tanks.

Liquids: Fluid Pressure/Surface Tension

Although it is not intuitively obvious, the buoyancy of the submarine does not change in any significant way, no matter how deep it goes. Water is essentially incompressible and its density is effectively the same at any depth. Drop an iron block into the ocean and it will continue to sink at its terminal velocity a concept that applies in liquids just as it does in gases all the way to the bottom. However, there is a limit to how deep a submarine can go, which is imposed by "water pressure", or the weight of the water above the submarine, which increases linearly with depth. Doubling the depth doubles the pressure.

There's nothing particularly tricky about this idea. Suppose Dexter has a set of identical concrete blocks, and proceed to stack them up one by one on top of each other. Of course, the force per unit area on the bottom of the stack grows linearly with each block that is added. There is no difference if he stacks up containers of water instead of concrete blocks, or for that matter adds uniform increments of water to a single tall container. What is not quite so intuitive is that the water pressure is the same in all directions. Suppose Dexter applies pressure to the top of a tank of water. Then, ignoring the increment of pressure added by the water in the tank as it gets deeper, the pressure of the water against the sides and bottom of the tank is exactly the same as the pressure on the water on top of the tank. One way to understand Pascal's principle is to understand the operation of a hydraulic jack. Suppose Dexter has a tank filled with water, with two tubes containing pistons fitted into the top. One tube has a cross-sectional area of 10 square centimeters, while the other has a cross-sectional area of 100 square centimeters.

If he exerts a downward force on the piston in the small tube, the transmission of water pressure on the piston in the big tube will result in an upward force ten times greater. In other words, this arrangement gives Dexter a mechanical advantage of 10:1. As with levers and screw jacks and so on, of course he doesn't get something for nothing. A simple consideration of the relative volume of water moved from the small tube to the big tube in this example shows that the piston in the big tube only moves a tenth of the distance upward that Dexter shoves the piston in the small tube down.

Another aspect of Pascal's principle is that it accounts for the buoyancy of a floating object, as discussed in the previous section. It is the force of pressure, exerted upward against the floating object, that prevents it from sinking further. Once the volume of water displaced equals the weight of the object, the force of water pressure will then support that object. In the definition of Pascal's principle above, the assumptions were that the surface of the tank was under heavy pressure and the tank was not very deep, meaning the pressure was the almost the same at the top and bottom of the tank and the difference could be ignored.

For water of any real depth, the increment of pressure caused by the accumulated weight of water as the depth increases is of course not negligible. Water has a mass of a tonne (1,000 kilograms) per cubic meter, and so on Earth, where the gravitational acceleration is 9.81 meters per second, a cubic tank of water a meter on a side exerts a force of 1,000 9.81 = 9,810 newtons on the square-meter base, or 9.81 kilopascals. If Dexter keeps the sides of the base the same size, but doubles the height to two meters, the pressure grows linearly to 19.6 kilopascals. If he goes up to three meters, the pressure goes up to 29.4 kilopascals.

The water pressure on the sides of the tank increases linearly from zero at the top to the same as the bottom when it reaches the bottom. However, let's suppose Dexter has a tank of water that's a meter high, a meter wide, and three meters long. This tank has a volume of three cubic meters. Is the bottom then under 29.4 kilopascals of pressure?

No, of course not. Pressure is force per unit area. He's increased the volume by a factor of three, but he's also increased the area of the base by a factor of three, and so the pressure remains the same as it was for the one-cubic-meter tank. The pressure is due to the weight of water pressing down only in the vertical direction. This means that it doesn't matter what size the base of the tank is. As long as the water in it is a meter deep, the pressure on the bottom in 9.8 kilopascals. This is more or less intuitive. What is not so intuitive is that, as dictated by Pascal's principle, water pressure is exerted in *all* directions, and so a dam has to be just as strong to hold back a body of water that extends, say, only 10 meters behind the dam, as it must be to hold back a body of water of equal depth that extends 100 kilometers behind the dam. The depth is the only real determinant of fluid pressure.

There are two basic forms of dams: "gravity" dams, which simply use great mass to hold back the water, and "arch" dams, which are built in narrow rocky canyons in the form of a horizontal arch, which transfers the massive load of the water to the canyon walls. Some dams may use both mass and an arched configuration to ensure structural stability. Obviously, the pressure of a liquid is equivalent to potential energy. If Dexter installs a valve at the bottom of a tank and opens it, then the water will pour out in a stream and will be able to do work, for example by turning a water-wheel.

This should not be surprising. A solid mass held at a height has potential energy due to its elevated position in a gravitational field, and the only difference between that and a liquid is that the liquid is in effect a continuous series of masses held at a continuous range of heights. This makes calculation of the precise value of potential energy of the liquid in a tank a bit tricky. We won't do that here, but the important thing is to realise that water pressure is equivalent to potential energy, in the same way that the tension of a compressed spring is equivalent to potential energy. Now let's suppose Dexter decides to tap this potential energy by opening the valve on the bottom of the tank. Water floods out in a stream. Since the water is now moving, it has kinetic energy. Of course, the higher the pressure at the valve, or equivalent the greater the height or "head" of water above the valve, the greater the kinetic energy of the flow and the greater its velocity. This is known as "Torricelli's principle", after the Italian physicist Evangelista Torricelli.

Since conservation of energy says that Dexter can't get something for nothing, this increase in kinetic energy must be obtained by a decrease in potential energy. As water pressure is equivalent to the potential energy of a liquid, this means that the pressure must decrease as well. In other words, the faster a liquid moves, somewhat counterintuitively the more its pressure decreases. However, it's not so different in concept from Dexter firing a ball out of a spring-loaded toy gun: the spring tension is released and converted into the kinetic energy of the ball.

The same effect is observed when water or another liquid passes from a pipe with a wide cross-section into a pipe with a narrow cross section. For a given rate of liquid flow, measured in liters per second, the velocity of the flow, measured in meters per second, has to increase if the pipe diameter decreases to maintain the flow. *In mathematical terms*: Liquid flow rate = Cross section1 × velocity1 = Cross section2 × velocity2.

The increased velocity of the liquid flow in the narrow section of pipe will result in a decrease in liquid pressure. The following illustration demonstrates this principle, which is known as the "Bernoulli effect", after the Swiss physicist and mathematician Daniel Bernoulli. On a small scale, liquids can demonstrate behaviour that would seem bizarre on a larger scale, due to attractions between the molecules of the liquid. For example, people can't walk on water, but insects like water-striders can. The intramolecular forces set up by water create "surface tension" that prevents a light object from simply falling right through the surface, even if it's denser than water.

Similarly, the adhesion of water to, say, glass surfaces will cause it to crawl up into glass tubes stuck into the surface of the water. This sort of "capillary action" is greater for tubes with very small diameters than it is for tubes with very large diameters. Again, this is due to a scaling effect, since doubling the diameter of a tube only doubles its circumference and the area of its glass walls, while the cross-sectional area and the amount of water that is to be lifted increases by four times.

One interesting application of these phenomena is the "restartable" rocket engine. A rocket engine that has to be restarted in space has to the deal with the problem that in free-fall the propellants won't flow down out of the tank to the engine, since there is no "down". There are various ways to deal with this problem; one ingenious scheme relevant to the current discussion is to fit the fuel tank with a set of internal grids of finer and finer spacing towards the bottom of the tank. The grids set up a "wicking" action that pulls the propellant to the bottom of the tank.

Gases

- Gases have much in common with liquids. They can flow, for example, and less dense objects will float. Gases also are subject to considerations of pressure like to those of a liquid; the weight of the atmosphere presses down on those who live at the bottom of it, and this atmospheric pressure drops off as we go to higher ground or fly up into the sky.

The atmospheric pressure of the Earth at sea level makes a convenient reference point for discussions of pressure, and one older measure of pressure is the "atmosphere", or 101,325 pascals. This is a somewhat arbitrary measure of pressure, however, and not convenient to work with using metric units, but the value of 100,000 pascals, known as a "bar", can be used as a reasonable approximation, all the more so because the pressure of air at sea level actually varies with changes in humidity, temperature, and wind conditions. A "barometer" is used to measure atmospheric pressure.

A traditional scheme for building a barometer involves a U-shaped tube partially filled with mercury. One end of the tube is sealed and the other is open to the atmosphere. Mercury is squirted into the sealed end of the tube while it's upside down, and when the tube is turned over, the mercury falls down to the bend in the tube, leaving behind a vacuum in the sealed side.

Air pressure will push the mercury up into the vacuum, and the height of mercury on the sealed side will provide an indication of atmospheric pressure. The mercury will be 76 centimeters higher in the sealed side of the tube than the open side of the tube at a standard atmosphere of pressure. Of course, since pressure is force per unit area, it doesn't matter how wide the tube is, as long as its diameter is constant. This scheme was discovered by Torricelli in 1643, and a now-disused measure of pressure named the "torr", equivalent to the displacement of a millimeter of mercury under standard conditions, was defined in his honour. Naturally, a standard atmosphere of pressure is equivalent to 760 torr, or 76 centimeters of mercury.

A more modern approach is to use a metal bellows that has been evacuated. The bellows is connected to an indicator needle through some mechanical arrangement that amplifies any minor motion in the bellows, and as the bellows flexes under changes in pressure, the needle moves to give the pressure value on the indicator scale.This device is known as an "anaeroid barometer". If the indicator scale is calibrated to give altitude instead of air pressure, the device is called an "altimeter". Manipulating the pressure of air or other gases is the basis of "pneumatic" technologies. The most basic pneumatic device is the hand-driven air pump or "bicycle pump" as it's sometimes called, which involves a cylinder with a piston that is moved up and down by muscle power.

There's two valves in the cylinder, an "input" valve that opens inward under a pressure difference and an "output" valve that opens outward under a pressure difference. Pulling up the handle on the pump creates lower pressure inside the cylinder, which causes the input valve to pop open, with ambient air entering to eliminate the pressure difference. Pushing down on the handle creates higher pressure inside the cylinder, forcing the input valve closed and popping the output valve open to push air into the tire or whatever.

The notion of the mercury barometer leads to the concept of the "siphon", in which a liquid such as water can be driven from a reservoir to an outlet through a pipeline over a hill or other obstruction. This vertical diversion is called a "siphon", and it works as long as two conditions are obeyed: the outlet has to be lower than the surface of the reservoir, or in other words water will still only flow downhill overall, and the siphon has to be filled with water before the scheme will work.

This looks like a problem in hydraulics, and it is, but the principle behind it is the same as that behind the mercury thermometer: the column of water flowing up the siphon is supported by the air pressure pressing down on the surface of the reservoir. When water flows out the outlet side of the siphon, it creates lower pressure, and so water on the inlet side of the siphon is driven upward. That's why the siphon has to be filled before the siphon effect will work: if the outlet side of the siphon was open to the air, the water level in the input side of the siphon would be exactly the same as it was in the reservoir.

There is a limit to the height or "crest" of the siphon above the reservoir level, depending on the density of the liquid and the air pressure above the reservoir. As mentioned, one atmosphere of pressure will lift mercury to a height of 76 centimeters, and since mercury is about 13.5 times more dense than water, one atmosphere of pressure on a reservoir should lift water about $13.5 \times 0.76 = 10.26$ meters.

This seems like a shallow crest for getting water over the top of a tall hill, but it must be remembered that the crest is relative to the water level in the reservoir, not relative to the bottom of the hill. Water reservoirs are often sited on high ground; the pipeline from the reservoir may drop down into a valley and then back up over a hill, but the crest of the hill will be about the same height or lower than the reservoir level.

Buoyancy issues are as relevant in gases as they are in liquids. For example, if a balloon is filled with a light gas such as hydrogen or helium, it will have a lower density than the ambient atmosphere and tend to float up. There is also the phenomenon of "lift" in heavier-than-air machines like aircraft and helicopters. There are misconceptions on how lift actually works, the biggest being that it is due to the Bernoulli effect, an error that actually pops up in elementary physics books.

As the story goes, the wings of an aircraft have a curved cross section, with a greater curvature on the top than on the bottom. Air flowing over the top of the wing has a longer path, and so its flow over the wing is more rapid. This means that the pressure on top of the wing is less than that on the bottom, and the effect is to provide a net upward force known as "lift" that keeps the machine flying. The problem with this scenario become immediately obvious when the question is asked: "So how can aircraft fly upside down?" They don't normally, since it's uncomfortable for the pilot, but any good flight demonstration team like the US Air Force Thunderbirds flies their jet fighters upside-down as a regular practice and the machines aren't immediately shoved into the ground.

The fallacy with the Bernoulli effect argument is that it assumes the airflow over the top of the wing has to maintain mass flow with the airflow over the bottom of the wing when it isn't constrained to do so. The Bernoulli effect may have some effect on lift, but it's not the primary mechanism for the phenomenon. The actual reason for lift is Newton's third law of motion. In flight, a wing is held at an "angle of attack (AOA)" upward of the direction of airflow, with the airflow striking the bottom surface and being deflected downward as "downwash".

The downward velocity component of the downwash provides a reaction to force the wing (and the aircraft connected to it) upward. The need to maintain AOA relative to the airflow does not mean the aircraft has to fly nose-up all the time; the wing may be fixed to the fuselage at a predetermined AOA relative to the aircraft's flight axis. As far as this description goes, a simple flat wing would work as well as any other, so why is a wing typically built with a curved upper surface? The trick is that such a wing would produce a good deal of turbulence over the top of the wing, causing "buffeting" as well as some "upwash" of the airflow, which would tend to force the aircraft back down.

Due to some subtle features of fluid flow discussed in a bit more detail below the airflow tends to hug the curve of the upper surface, which means that it leaves the trailing edge of the wing in a downward direction, also contributing to lift. The higher the AOA of the wing, incidentally, the greater the drag, and so there's a tradeoff between AOA and speed. It is not unusual for aircraft to have a landing gear configuration that leaves them nose-high, increasing the AOA until the aircraft gets off the ground and starts flying level. A high AOA is particularly important for carrier takeoffs and landings, since the aircraft has to get up or down in a short space.

A number of British carrier jets of the 1950s and 1960s were actually launched off carrier catapults strapped down so their nosewheel was in the air to ensure a high AOA. The US Navy Vought F-8 Crusader fighter had a wing that hinged upward for takeoffs and landings to permit a high effective AOA while giving the pilot a good view forward. Such considerations of lift show that a conventional aircraft, even though it can in principle fly upside down, isn't as efficient in flight when doing so, having to maintain a nose-up attitude to ensure a usable AOA. Stunter aircraft may have no fixed AOA and wings that are curved on both sides.

An aircraft has "flight control surfaces" that modify the airflow to control the machine's direction. There are moving panels at the trailing edge of each wing known as "ailerons" that are pivoted in an alternating fashion on one side up, on the other down to allow the aircraft to "roll". There are "elevators" at the trailing edge of the horizontal tailplane that are pivoted together to cause the aircraft to "pitch" up and down. Finally, there is a "rudder" at the trailing edge of the vertical tailplane that can be pivoted to cause the aircraft to "yaw" back and forth. These are the "basic" control surfaces, but there are many variations. For example, many tailless aircraft, usually delta-winged machines, have an "elevon" on each wing, with both moved up or down to provide elevator pitch control, or moved in opposite directions to provide roll control. Tailless flying wing aircraft that don't have a tailfin, such as the B-2 Stealth bomber, have "rudderons" that are split top and bottom on the outer rear edge of the wing.

Both rudderons can be opened up top and bottom on one side to cause the aircraft to turn, or can be opened up top and bottom on both sides of the aircraft to act as airbrakes. An elevon is fitted on each half of the wing inboard of the rudderon to provide pitch and roll control. Incidentally, moving the rudderons away from their neutral position greatly increases the B-2's visibility to radar, so when it is flying into a combat area it steers by adjusting engine thrust, though this is more troublesome and less effective than using rudderons.

Aircraft may also have a number of auxiliary control surfaces. "Flaps" can be extended below the wings to increase the downwash and lift for take-off or low-speed flight, with flaps cascaded in some cases to increase their length. Control surfaces named "spoilers" or "lift dumpers" can be attached to the top of the wing to disrupt airflow and destroy lift, helping the aircraft to land. High-altitude aircraft have very wide wingspans and tend to be reluctant to come down on the runway, so their wings are usually fitted with lift dumpers.

Like a submarine, an aircraft has to maintain trim. This is obvious in the case of cargoes, which clearly can't be loaded onto a cargo plane in an unbalanced fashion, and the crew of a large cargo aircraft will generally include a specialist called a "loadmaster" who ensures that the cargoes are balanced properly incidentally, maritime cargo vessels also have loadmasters to reduce the hazard of capsizing due to an unbalanced load. Aircraft trim is less obvious in the important case of fuel management.

Usually aircraft designers will try to put fuel tanks at the centre of gravity of an aircraft, but an aircraft may have several fuel tanks, and care often must be taken to empty the tanks in a specific order, or the aircraft may become unbalanced. This was done manually by the pilot in the past, but now sophisticated aircraft have automatic systems to ensure the proper sequencing. Since an aircraft will always have slight imbalances, instead of having the pilot try to compensate for trim faults by keeping the control surfaces slightly activated at all times, which would be very tiring, traditionally aircraft have featured small control surfaces called "trim tabs" that can be set to a fixed position using a dashboard knob to keep the machine flying level. Some aircraft may instead use knobs to introduce a slight bias in the aircraft's flight control surfaces.

Complications: Viscous Flows and Compressible Flows

The discussion to this point on the flows of liquids and gases which can be lumped together as "fluids", meaning both can flow has been simplified by assuming the flows aren't viscous that is, subject to friction forces and are incompressible. These are reasonable assumptions for a simple analysis, but the real world can be substantially more complicated. Formal studies of the effects of viscosity on fluid flow began in the middle of the 19th century, leading to the invention of equations for describing the viscous flow of an incompressible liquid; the equations were developed independently by Claude Louis Marie Navier, a French engineer, and Sir George Gabriel Stokes, a British mathematician, and so these rules are known as the "Navier-Stokes equations".

They're not simple equations, and in fact they are only fully solvable for simple scenarios. Bernoulli's principle no longer strictly applies since the viscous friction dissipates energy of the flow, resulting in a pressure drop along the pipe. At the outset, it was thought that this pressure drop would increase in direct proportion to the flow velocity that is, double the flow velocity and the pressure drop doubles as well but experiments showed that this was only true up to a certain threshold. Beyond that threshold, the pressure drop increased with roughly the square of the flow velocity. This hinted that there were two different processes at work, which was demonstrated by Osborne Reynolds, a professor at the University of Manchester in Britain. Osborne devised an apparatus in which he could observe the flow of dye in water pumped through an tube.

He observed that the dye would flow straight for a distance, what was called "laminar" flow, and then break up into a tangle of eddies, what was called "turbulent" flow. This phenomenon can be observed by watching smoke float up off the tip of a lit cigarette: the smoke will rise for a space and then break into a tumble of coils. Incidentally, turbulent fluid flows are an example of a chaotic system. In any case, the drag of the flow increases dramatically when the flow changes from laminar to turbulent. Reynolds was able to devise an equation that factored in the flow velocity, the density of the fluid, the viscosity of the fluid, and the length of fluid flow to return a value that would indicate when the flow changed from laminar to turbulent. The

"Reynolds equation" is given as follows: Fluid velocity × length of fluid flow × fluid density/fluid viscosity

The value returned by the Reynolds equation is known, to no great surprise, as the "Reynolds number (RN)". If the RN is less than 2,100, the flow will always be laminar. It doesn't necessarily switch to turbulent just above that value, but it becomes increasingly unstable and likely to shift to turbulent as the RN increases beyond 2,100. One of the interesting features of the RN is that fluid flows with similar RNs behave in much the same way, no matter if the fluid is air or water or whatever. A sailboat moving at low speed in a dense fluid like water has much the same behaviour in terms of fluid dynamics as an aircraft moving much faster in a thin fluid like air. This has a major, though not obvious, implication for issues of scaling in aircraft design.

The smaller an aircraft gets, the greater the effective viscosity of the air. To a gnat, the air feels thick and viscous, and flying is more like swimming in fact, the smaller the creature, the less bother it is to get into the air, and small spiders will sometimes migrate by simply throwing out a thread of silk to the wind. Researchers working on tiny robot aircraft have finding out that conventional aircraft designs are not very workable at such small scale, and they have gravitated towards unusual designs, such as circular flying wings, or even "ornithopters" that flap their wings to fly. The complexity of the analysis of viscous flows meant that scientists and engineers really needed to obtain simplified but still workable methods. In 1904, the German engineer Ludwig Prandtl observed that flows could be often be separated into two regions: a thin region known as the "boundary layer" close to the surface over which the flow passes and where the viscous effects are concentrated, and the region of flow above the boundary layer where viscous effects could be disregarded.

It is this viscous boundary layer that causes the airflow over the top of a wing to follow the curve of the wing in effect, it causes the airflow to stick to the wing. At low speeds, the boundary layer will tend to "separate" from the wing and ruin lift, and so a number of combat jets were built with a "blown flaps" or "boundary lsayer control (BLC)" scheme in which jet engine bleed air was blown over the top of the wing or the flaps to ensure that low-speed separation didn't occur.

The boundary layer can also be a nuisance for combat jet aircraft. The engines of such aircraft require a continuous large airflow to provide a high level of thrust, and if the engine intakes are close to the aircraft's fuselage they may swallow the viscous ("stagnant")

boundary-layer air, choking off airflow and engine power even leading to "engine stall" that can "flame out" the engine or, in the worst case, damage it. Such aircraft often have a "splitter plate" between the front of each engine intake and the fuselage; these plates are set slightly off from the fuselage and prevent the boundary-layer air from being sucked into the inlet.

Assuming that a fluid is incompressible is appropriate if the fluid is a liquid, but it can lead to drastically wrong conclusions when the fluid is a gas. Even assuming nonviscous laminar flow, Bernoulli's law only really applies to slow-moving gas flows; if the gas flows are rapid, compression comes into play, leading to distinctly different behaviour.

Analysis of a compressible flow is too tricky to be discussed in detail in this document, but a simple example will illustrate the principles involved. Consider a standard rocket nozzle: it can be modeled as a cylinder that leads to a constriction, known as the "throat", which leads into a widening "exhaust bell" open at the end. Hot gases flowing down the hit the throat, where the velocity picks up and the pressure falls, roughly as it would in a fluid, though compression occurs as well. Once the hot gases exit the throat and enter the widening exhaust bell, they expand rapidly, and this expansion drives the velocity up while the pressure continues to fall exactly the reverse scenario of what would occur if a fluid passed through a constriction into a widening bell.

A standard rocket nozzle is designed to provide optimum thrust for a certain range of altitudes and atmospheric pressures. High-performance jet combat aircraft must operate effectively over a wide range of conditions, and so they have exhausts whose geometry can be varied depending on need.A fully "authentic" computer simulation of a real-life gaseous flow that takes into consideration viscous effects, turbulence, and compression can require some serious computing horsepower.There is a complicated interaction between all the facets of the simulation which can get even more complicated if chemical transformations take place in the stream. One aspect to the complexity is that the pressure, volume, and temperature of a gas are closely related to each other. This leads to a consideration of concepts of "temperature" and "heat", which are somewhat devious to define.

Although this document has mentioned temperature effects i`n a casual fashion, to go much further requires a more detailed discussion. Similarly, although matter can often be converted between solid, liquid, and gaseous forms through changes in temperature, that also requires a more detailed look.

Speculative Atomism

Actual proof of the reality of atoms is an achievement of this century; even at its beginning critical consideration required the confession that atoms, although investigators spoke of them, were still no more than a stimulating and useful hypothesis, which in the final analysis was unproved and possibly unnecessary, perhaps even misleading. Although the idea of atomism was important and fruitful at the very beginning of western natural research, this idea did not originate from western research itself; just as we borrowed the model for mathematical thinking from the Greeks, we took the atomistic interpretation of nature from them. Nothing exists—Democritus taught—except atoms and vacuum; everything else is imagination. These innumerable atoms, indestructible and invariable elementary constituents, are to be considered the basis for all beings and events in nature.

Individual atoms possess invariable geometric form and motions which fluctuate due to the pressure of and collision with other atoms. In all changes in the structure of nature the atoms are always preserved; from nothing comes nothing; nothing which exists can be destroyed; change is merely combination and separation of parts. The differences of all things stem from the diversity of their atoms in number, magnitude, structure and arrangement. Since the movements of each individual atom are regulated according to natural law, nothing arbitrary can happen in all nature; nothing happens accidentally, but everything from a cause and of necessity.

Our rough senses can not recognize these amazingly fine elements in their true nature and form, but experience their effects only vaguely. "Only in the imagination does sweetness exist, in the imagination bitterness, in the imagination heat, cold, colours; in reality nothing exists but atoms and vacuum." This is marked evidence of that "decoloration" of the world, the assurance that all direct sensory observation is an illusion and that a knowledge of the true status of things must lead us to a picture in which nothing remains but the geometrical form and mechanical motion of the atoms.

These idea forms of materialistic philosophy introduce the possibility of considering all events in nature as results of strict regularity; the gist of the basic concept of natural scientific thought is anticipated here in general and is thoroughly formed for the first time. On the other hand we must not forget to recognize how radically this philosophy opposes religious representations of its time. For the Greeks every stream was a God; in the springs lived the nymphs, in the woods the horned Pan, and in caves and caverns resided demons. This entire mythological picture of the world, which suspected the arbitrary and incalculable influence of demoniacal powers in every natural process, was pushed aside by the powerful conception of a picture of nature of stricter regularity. Epicurus, who renewed Democritus' teaching, and who in his thinking and conduct of life combined a rationalistic, vigorous attitude with a very cultivated, fine-spirited being, was not so brutal as to deny existence completely to the Gods. He let them continue as blessed, immortal beings, occupied with themselves, and never infringing upon the workings of the world—these workings being developed according to the mechanical lawfulness of atomic motions in strict, definite sequence. In a great didactic poem *"De rerum natura"* the Roman Lucretius explained completely the ideas of atomistic regularity for all the fields of nature known at that time. These teachings were made known to western scholars in connection with the humanistic studies of the Renaissance; Gassendi, principally, introduced Greek philosophy into western science.

Let us consider the ideas which Newton, together with Gassendi, applying the divergent theories of Descartes, formed about the construction of matter. F. Dannemann explains them as follows in his historical work [1] on the natural sciences: "He considered it most plausible, that it (matter) consists of solid, impermeable, moving particles. Since natural bodies, *e.g.*, water, are invariable in their properties, the particles of which they consist must neither be able to be used up nor destroyed. The variation of material things is to be laid exclusively to the separations, combinations and movements of those invariable particles." These are, as is obvious, exactly the concepts of materialistic philosophy, almost unchanged, just as Democritus stated them. But Newton was—outside of natural science—in no wise an adherent of materialistic theories. For him these ideas meant nothing but a clue to physical research; he did not see in them the support of radical, world-viewing results.

Chapter 2

Models of Atom

Atomic Spectra

The first person to realise that white light was made up of the colours of the rainbow was Isaac Newton, who in 1666 passed sunlight through a narrow slit, then a prism, to project the colored spectrum on to a wall. This effect had been noticed previously, of course, not least in the sky, but previous attempts to explain it, by Descartes and others, had suggested that the white light became colored when it was refracted, the colour depending on the angle of refraction. Newton clarified the situation by using a second prism to reconstitute the white light, making much more plausible the idea that the white light was composed of the separate colours. He then took a monochromatic component from the spectrum generated by one prism and passed it through a second prism, establishing that no further colours were generated. That is, light of a single colour did not change colour on refraction. He concluded that white light was made up of all the colours of the rainbow, and that on passing through a prism, these different colours were refracted through slightly different angles, thus separating them into the observed spectrum.

In 1752, the Scottish physicist Thomas Melvill discovered that putting different substances in flames, and passing the light through a prism, gave differently patterned spectra. Ordinary table salt, for example, generated a "bright yellow". Furthermore, not all the colours of the rainbow appeared - there were dark gaps in the spectrum, in fact for some materials there were just a few patches of light. By the 1820's, Herschel had recognized that spectra provided an excellent way to detect and identify small quantities of an element in a powder put into a flame. Meanwhile, the white light of the sun was coming in for more detailed scrutiny. In 1802, William Wollaston in England had discovered (perhaps by using a thinner slit or a better prism) that in fact the solar spectrum itself had tiny gaps - there were many thin dark lines in the rainbow of colours.

These were investigated much more systematically by Joseph von Fraunhofer, beginning in 1814. He increased the dispersion by using more than one prism. He found an "almost countless number" of lines. He labeled the strongest dark lines A, B, C, D, etc.

Foucault Connects Melvill's Bright Lines and Fraunhofer's Dark Lines

In 1849, Foucault (of speed of light and pendulum fame) examined the spectrum of light from a voltaic arc between carbon poles. He saw a bright double yellow line at exactly the same wavelength as Fraunhofer's dark D line in the solar spectrum. Investigating further, Foucault passed the sun's light through the arc, then through a prism. He observed that the D lines in the spectrum were even darker than usual. After testing with other sources, he concluded that the arc, which emitted light at the D line frequency, would also absorb light from another source at that frequency.

This discovery did not surprise Sir George Stokes in Cambridge. He pointed out that any mechanical system with a natural frequency of oscillation will emit at that frequency if disturbed, but will also absorb most readily at that frequency from incoming disturbances, the phenomenon of resonance.

Question: In a total eclipse of the sun, the only sunlight reaching earth comes from the hot gases of the sun's atmosphere, light from the sun's main disc being blocked by our moon. The light from these hot gases was analysed during an eclipse in 1870. How do you think the spectrum observed related to that of full sunlight? The spectrum of *hydrogen*, which turned out to be crucial in providing the first insight into atomic structure over half a century later, was first observed by Anders Angstrom in Uppsala, Sweden, in 1853. His communication was translated into English in 1855. Angstrom, the son of a country minister, was a reserved person, not interested in the social life that centered around the court.

Consequently, it was many years before his achievements were recognized, at home or abroad (most of his results were published in Swedish). Meanwhile, in Freeport, Pennsylvania, in 1855, David Alter described the spectrum of hydrogen and other gases.

In the 1840's, Alter had started the first commercial production of bromine from brines. He also found a way to extract oil from coal, but that proved uneconomic after the discovery of oil in Pennsylvania. His work was not widely recognized, either.

Bunsen and Kirchhoff

The first really systematic investigation of spectra was that of Bunsen and Kirchhoff, in Heidelberg, between 1855 and 1863. They used several techniques. For one thing, they introduced various salts intowhat else?the flame of a Bunsen burner. This was a very effective way of viewing spectra, because the Bunsen burner flame itself gave out practically no light. They also used the cooler flame of alcohol burning mixed with water to generate a vapour to study absorption spectra. Finally, they studied the spectra of electric arcs between electrodes of different materials.

Using iron electrodes gave a spectrum that coincided with dark lines in the sun's spectrum. Copper electrodes did not. They concluded that the sun's atmosphere contained iron, but not much copper, and that, they said, seemed very plausible since there is so much iron in the earth, and in meteors. (*Cautionary note to philosophers*: In 1835, the French philosopher Auguste Comte (the founder of positivism) wrote: "...Our knowledge concerning the gaseous envelopes [of stars] is necessarily limited to their existence, size and refractive power, we shall not at all be able to determine their chemical composition or even their density. The collaboration of Kirchhoff and Bunsen was a major research effort, even by modern standards. They determined thousands of spectral lines, each to an accuracy of one part in ten thousand. They spectroscopically discovered new elements: rubidium and cesium.

Their method was used to find fifteen more new elements before the end of the century. In 1869, Joseph Lockyer studied the spectra of solar prominences (in eclipses). He found the spectra to be slightly Doppler shifted, so was able to deduce the speeds of the gases whirling around the sunspots. He also found a spectrum never seen before, and conjectured that it came from a new element he named Helium. In fact, helium was later discovered on earth in 1895, by Ramsay. At that time, it had just become evident that there was an inert component, argon, in the earth's atmosphere. Earlier, an inert gas had been observed to emanate from uranium salts when they were heated.

Ramsay assumed this would be the same gas, but decided to check. On heating uranium salts and performing a spectral analysis of the emitted gas, much to his surprise he found it to be helium.

The Balmer Series

It is clear from the above that a tremendous amount of scientific progress was made using spectral lines, yet no-one had the slightest idea why atoms emitted at the frequencies they did. It was appreciated that spectra implied that atoms had structure.

In 1852, Stokes had stated that probably the vibrations that produced light were vibrations among the constituent parts of molecules (a term which also included atoms at that time) and in 1875 Maxwell, in enumerating properties atoms must have, included the capability of internal motion or vibration. This worried Maxwell, though. As he said, the spectroscopic evidence forces the conclusion that the atom is quite complex, with many internal degrees of freedom. Yet apparently all these modes of vibration, or almost all of them, are not excited by heat, since if they were this extra capacity of the atom to absorb energy would be reflected in its specific heat. Obviously, if any pattern could be discerned in the spectral lines for an atom, that might be a clue as to the internal structure of the atom.

One might be able to build a model. A great deal of effort went into analysing the spectral data from the 1860's on. The big breakthrough was made by Johann Balmer, a math and Latin teacher at a girls' school in Basel, Switzerland. Balmer had done no physics before, and made his great discovery when he was almost sixty. He decided that the most likely atom to show simple spectral patterns was the lightest atom, hydrogen. Angstrom had measured the four visible spectral lines to have wavelengths 6562.10, 4860.74, 4340.1 and 4101.2 in Angstrom units (10^{-10} meters). Balmer concentrated on just these four numbers, and found they were given by the formula:

$$l = b\left(\frac{n^2}{n^2-4}\right)$$

where b = 3645.6 Angstroms, and n = 3, 4, 5, 6. Balmer suggested that there would be other lines - in the infrared - corresponding to n = 7, 8, etc., and in fact some of them had already been observed, unbeknownst to Balmer. He further conjectured that the 4 could be replaced by 9, 16, 25, ... and this also turned out to be true - but these lines, further into the infrared, were not detected until the early twentieth century, along with the ultraviolet lines generated by replacing the 4 by 1.

It is instructive to write Balmer's general formula in terms of the inverse wavelength. This is called the wave number - the number of waves that fit in one unit of length.

$$\frac{1}{\lambda} = R_H = \left(\frac{1}{n^2} - \frac{1}{m^2}\right)$$

where n, m are integers, and R_H is the Rydberg constant, 109,737 cm^{-1}. This constant is named after the Swedish physicist Rydberg who (in 1888) presented a generalization of Balmer's formula, in which the integer n was replaced by n + constant, the constant being less than unity. Rydberg suggested that all atomic spectra formed families with this pattern. (He also said he was unaware of Balmer's work.) It turns out that there *are* families of spectra following Rydberg's pattern, notably in the alkali metals, sodium, potassium, etc., but not with the precision the hydrogen atom lines fit the Balmer formula, and low values of n give lines that deviate considerably.

(Modern footnote: atoms having spectral lines following Rydberg's formula are called Rydberg atoms. These Rydberg atoms have one electron orbiting at a much greater distance from the nucleus than the others. Consequently, Rydberg atoms can only survive in a gas at very low pressure, otherwise that outermost electron gets knocked off. Prof. Tom Gallagher in our Department is a world expert on these atoms, which have proven a rich source of information on the quantum mechanics of atomic structure.)

One pattern that was noticed in the spectra of many atoms is Ritz' Combination Principle: if for a given atom there are spectral lines at two wave numbers, there is sometimes another spectral line at the precise sum of those two wave numbers.It is easy to see from Balmer's formula that this is true for some pairs of lines in the hydrogen spectrum. It also turns out to be true for atoms where the spectral lines have no other discernible pattern.

Models of the Atom

The first attempt to construct a physical model of an atom was made by William Thomson (later elevated to Lord Kelvin) in 1867. The most striking property of the atom was its *permanence*. It was difficult to imagine any small solid entity that could not be broken, given the right force, temperature or chemical reaction.

In contemplating what kinds of physical systems exhibited permanence, Thomson was inspired by a paper Helmholtz had written in 1858 on *vortices*. This work had been translated into English by a Scotsman, Peter Tait, who showed Thomson some ingenious experiments with smoke rings to illustrate Helmholtz' ideas. The main point was that in an *ideal* fluid, a vortex line is always composed of the same particles, it remains *unbroken*, so it is ring-like. Vortices can also form interesting combinationsA good demonstration is provided by creating two vortex rings one right after the other going in the same direction. They can trap each other, each going through the other in succession. This is probably what Tait showed Thomson, and it gave Thomson the idea that atoms might somehow be vortices in the ether. Of course, in a non ideal fluid like air, the vortices dissipate after a while, so Helholtz' mathematical theorem about their permanence is only approximate.

But Thomson was excited because the ether *was* thought an ideal fluid, so vortices in the ether might last forever! This was very aesthetically appealing to everybody - "Kirchhoff, a man of cold temperament, can be roused to enthusiasm when speaking of it."

In fact, the investigations of vortices, trying to match their properties with those of atoms, led to a much better understanding of the hydrodynamics of vortices - the constancy of the circulation around a vortex, for example, is known as Kelvin's law.

In 1882 another Thomson, J. J., won a prize for an essay on vortex atoms, and how they might interact chemically. After that, though, interest began to wane - Kelvin himself began to doubt that his model really had much to do with atoms, and when the electron was discovered by J. J. in 1897, and was clearly a component of all atoms, different kinds of non-vortex atomic models evolved. It is fascinating to note that the most exciting theory of fundamental particles at the present time, *string* theory, has a definite resemblance to Thomson's vortex atoms. One of the basic entities is the closed string, a little loop, which has fields flowing around it reminiscent of the swirl of ethereal fluid in Thomson's atom. And it's a very beautiful theory - Kirchhoff would have been enthusiastic!

Floating Magnets

In 1878, Alfred Mayer, at the University of Maryland, dreamed up a neat demonstration of how he imagined atoms might be arranged in molecules. He took a few equally magnetized needles and stuck them through corks so that they would float with their north poles all at the same height above the water, all repelling each other equally.

He then held the south pole of a more powerful magnet some distance above the water, to attract the needles towards this central point. The idea was to see what equilibrium patterns the needles would form for different numbers of needles. He found something remarkable - the needles liked to arrange themselves in shells. Three to five magnets just formed a triangle, square and pentagon in succession. but for six magnets, one went to the centre and the others formed a pentagon. For more magnets, an outer shell began to form. Kelvin's immediate response to Mayer's publication was that this should give some clues about the vortex atom. Apparently it didn't, but twenty-five years later it guided his thinking on a new model.

Plum Pudding

Kelvin, in 1903, proposed that the atom have the newly discovered electrons embedded somehow in a sphere of uniform positive charge, this sphere being the full size of the atom. (Of course, the sphere itself must be held together by unknown non-electrical forces - which is still true of the positive charge in our modern model of the atom.)

This picture was taken up by J. J. Thomson too, and was dubbed the plum pudding model, after traditional English Christmas fare, a large round pudding (rich with suet) with raisins embedded in it.

In 1906, J. J. concluded from an analysis of the scattering of X-rays by gases and of absorption of beta-rays by solids, both of which he assumed were effected by electrons, that the number of electrons in an atom was approximately equal to the atomic number. This led to a picture of electron arrangements in an atom reminiscent of Mayer's magnets. Perhaps by analysing possible modes of vibration of electrons in these configurations, the spectra could be calculated. The simplest case to consider was clearly hydrogen, now assumed (correctly) to contain just one electron.

How Does an Atom's Colour Depend on its Size?

By "colour" we mean here the spectral colours emitted when the atom is excited. In Thomson's plum pudding model, there is a clear relationship between the *size* of the pudding and the *frequency* at which the electron will oscillate, and hence presumably radiate, when excited. The two are related because the assumption is that the total positive charge - which is uniformly spread throughout the sphere is just equal to the electron's negative charge. At rest in its lowest state, the electron just sits in the middle of this sphere of charge. When bumped somehow, it will oscillate about that point.

If the electron is at distance x from the centre, it will feel a restoring force towards the centre equal to the attraction from that part of the positive charge it is "outside" of - that is, the charge within a sphere of radius x about the centre. Therefore, the *larger* the whole atomthe pudding - the more thinly spread the positive charge is, and the *smaller* the amount of charge within the small sphere of radius

x that is attracting the electron back towards the centre. So, the bigger the atom is, the slower the electron's oscillation is, and the lower frequency the radiation emitted. It is straightforward to give a quantitative estimate of the size of the atom based on the observation that when excited it emits radiation in the visible range. Let us assume that the positively charged sphere has radius r_0 (this is then the size of the atom, which we know is about 10^{-10} meters).

If the electron is displaced from the centre of the atom in the x-direction an amount x, it is attracted back by all the charge that is now closer to the centre than itself, that is, an amount of charge equal to ex^3/r_0^3. (Recall e is the total amount of charge on the sphere, and x^3/r_0^3 is the fraction of the sphere closer to the centre than x.) This charge acts as if it were a point charge at the origin, so the inverse-square law gives a $1/x^2$ factor, and the equation of motion for the electron is therefore:

$$m\frac{d^2x}{dt^2} = -\frac{1}{4\pi\varepsilon_0}\cdot\frac{e^2x}{r_0^3}$$

Provided it stays within the sphere, the electron will execute simple harmonic motion with a frequency

$$w^2 = \frac{1}{4\pi\varepsilon_0}\cdot\frac{e^2}{mr_0^3}.$$

Notice that, as we discussed above, as the size of the atom increases the frequency goes down. And we know the frequency, at least approximately it corresponds to visible light.

Therefore, this model will predict a size of the atom, which we can compare with the size from other predictions, such as Brownian motion (plus the assumption that in a liquid, the atoms are fairly close packed - they take up most of the room available). If we take visible light, say with a frequency 4.10^{15} radians per second, we find r_0 must be about 2.10^{-10} meters, a little on the large side, but encouragingly close to the right answer for a first attempt. Sad to report, though, no real progress was made beyond this in predicting spectra using Thomson's pudding.

Many attempts were made to find stable arrangements of electrons in atoms, not just hydrogen, using models like Mayer's magnets, and also having the electrons going around in circles. It was hoped that if certain numbers of magnets formed a very stable arrangement, that might model a chemically nonreactive atom, etc. - but nobody succeeded in making any real predictions along these lines, the models could not be connected with the properties of real atoms. Evidently, then, the theorists were stuck - and the experimental challenge was to find some way to look *inside* an atom, and see how the electrons were arranged.

The Bohr Atom

In 1911, the 26-year-old Niels Bohr earned a Ph. D. at the University of Copenhagen; his dissertation was titled "Studies on the Electron Theory of Metals". He was awarded a postdoctoral fellowship funded by the Carlsberg Brewery Foundation, which enabled him to go to Cambridge in September to study with J. J. Thomson. Bohr was a great admirer of Thomson's many achievements, both experimental and theoretical.

In his thesis work, he had closely studied some of the problems covered in Thomson's book *Conduction of Electricity through Gases.* He had uncovered some apparent errors in Thomson's work, and looked forward to discussing these points with the great man. Unfortunately, by the time Bohr arrived, the Cavendish Laboratory had grown to the point where Thomson as director had more than he could manage. He had no spare time to think about electrons, and was not happy to hear from Bohr that some of his earlier work might be incorrect.

In fact, Thomson went out of his way to avoid theoretical discussions with Bohr. He did assign Bohr an experiment on positive rays, but Bohr was not enthusiastic. Bohr kept himself busy writing a paper on electrons in metals, reading Dickens to improve his English, and playing soccer.

In December, Rutherford came down from Manchester for the annual Cavendish dinner. Bohr later said that he was deeply impressed by Rutherford's charm, his force of personality, and his patience to listen to every young man who might have an ideacertainly a refreshing change after J. J.! A little later, Bohr met with Rutherford again when he visited one of his father's friends in Manchester, someone who also knew Rutherford.

Although Rutherford was usually skeptical of theorists, he liked Bohr. For one thing, Rutherford was a soccer fan, and Bohr's brother Harald (only nineteen months younger than Bohr) was famoushe had played in the silver medal winning Danish soccer team at the 1908 Olympics in London. After talking it over with Harald, who visited Cambridge in January, Bohr moved to Manchester in March, and took a six-week lab course, given by Geiger, Marsden and others. Really, though, his interests were theoretical, and he talked a lot with Charles Galton Darwin"grandson of the real Darwin", as Bohr put it in a letter to Harald.

Darwin had just completed a theoretical analysis of the loss of energy of an a -particle going through matterthat is, an a that doesn't get close enough to a nucleus to be scattered. Such a 's gradually lose energy by churning through the electrons, and the rate of loss depends on how many electrons they encounter.

In particular, Bohr concluded, after reviewing and improving on Darwin's work, it seemed clear that the hydrogen atom almost certainly had a *single* electron outside the nucleus.

What Determines Atomic Size?

A big problem with the nuclear hydrogen atom was: what determined its size? Classical mechanics gives a simple dynamical equation for circular orbits:

$$\frac{mv^2}{r} = \frac{1}{4\pi\varepsilon_0} \cdot \frac{e^2}{r^2}.$$

Now this equation is satisfied by *any* circular orbit centered at the nucleus, however large or small. (Note, by the way, that multiplying both sides by $r/2$ gives that the magnitude of the kinetic energy in the circular orbit is just half the magnitude of the negative potential energy. We need this below.)

There is no hint here that the atom in its "natural" ground state should have any particular radius. But it does! This means we're missing *something*. But what? Bohr (and others) thought that Planck's constant must somehow play a role in determining the size of the orbit. After all, it *did* play a role in restricting allowed orbital changes in the oscillators in black body radiationand these oscillators, although not very clearly understood, were of the same general size as atoms. So evidently the standard picture of how an oscillating charge radiated couldn't be right at the atomic level.

Bohr concluded that in an atom in its natural rest state, the electron must be in a special orbit, he called it a "stationary state" to which the usual rules of electromagnetic radiation didn't apply.

In this orbit, which determined the size of the atom, the electron, mysteriously, didn't radiate. Just how to bring Planck's constant into a discussion of the hydrogen atom was not so clear, though. For the black body oscillators, it related the frequency f of the oscillator with the allowed energy change E by $E = hf$. The obvious parallel approach for the hydrogen atom was to identify the frequency f with the circular frequency of the electron in its orbit.

However, in contrast to the simple harmonic oscillator this hydrogen atom frequency *varied* with the size of the orbit. Still, it was the only frequency around, and, dimensionally, multiplying it by h gave an energy. What energy could that be identified with? Again, the choice was limitedthe electron had a kinetic energy E, the potential energy was $-2E$ and the total energy $-E$.

If a hydrogen nucleus captured a passing electron into its ground state, and emitted one quantum of electromagnetic radiation, that quantum would have energy E, the same as the electron kinetic energy in the natural stationary state (called the *ground state*).

Bohr suggested in a note to Rutherford in the summer of 1912 that requiring this energy be some constant (assumed to be of order of magnitude one) multiplied by hf would fix the size of the atom. Actually his argument was a bit more complicated, he considered the several electron atom, and took the electrons to form rings. However, the basic point is the samea condition like this constrains the atomic size, it would be fixed uniquely if we knew the constant. If we assume the constant is 1, for example, we have

$$\frac{1}{2}mv^2 = \text{hv/2pr}$$

Putting this together with the dynamic equation above determines the atomic radius r. It is easy to check that it predicts a radius of $4\text{p e}\,_0h^2/\text{p}\,^2me^2$, which is just four times the "right answer" defined as the Bohr radius. The correct Bohr radius comes out if we choose the constant to be one-half, $E = ½hf$, which Bohr used later.

Hence the approximate size of the atom follows from *dimensional* arguments alone once one assumes that Planck's constant plays a role! Of course, the nucleus is irrelevant in determining the atomic sizeit just provides a fixed centre of electrostatic attraction. The relevant electronic parameters are the mass m and the strength of attraction $e^2/4\text{p e}\,_0$. Together with h, these parameters determine a length.

It should be mentioned that this assumption explained more than the size of the hydrogen atom. It was believed at the time that in the higher atoms, the electrons formed rings, thought to lie one outside the other, and various stability arguments indicated that there couldn't be more than seven electrons in a ring.

The length scale above, $4\pi\varepsilon_0 h^2/2\pi^2 me^2$, would *decrease* for larger atoms, with e^2 replaced by Ze^2 essentially, for nuclear charge Z. Thus as the number of rings increased, the size of the rings would decrease, explaining the observed approximate periodicity in atomic volume with atomic number.

Also, in 1911 Richard Whiddington in Cambridge had found that to cause a substance having atomic number *A* to emit characteristic x-rays by bombarding it with electrons, it was necessary to use electrons of speed approximately $A \times 10^6$ meters per second.

Any substance on being bombarded with sufficiently fast electrons emits a continuum of x-ray frequencies up to a maximum frequency f given by hf = kinetic energy of electron, *plus* some sharply defined linesx-rays at a particular frequency, which does not change as the electron speed is further increased.

The frequency corresponding to these lines was found to increase with atomic number. Applying his length scale argument to the innermost ring of an atom, Bohr found that an electron in that ring would have a speed proportional to the nuclear charge, and hence, at least approximately, to the atomic number. Furthermore, the predicted speed in orbit was of the same order as that of Whiddington's electrons.

Nicolson: a Clever Idea about a Wrong Model

Meanwhile, in Cambridge one J. W. Nicolson was struggling to incorporate Planck's ideas in a model of the atom, in an attempt to understand some strange sets of spectral lines observed in nebulae and in the sun's corona.

He conceived a rather exotic (and quite wrong!) model, in which a ring of electrons, like a necklace, orbited the nucleus. (Actually, many people, including Bohr himself, investigated models like this.

The reason was that the classical radiation from a ring of electrons is a lot less that that from a single orbiting electron, the fields tend to cancel each other.) Oscillations of electrons in this ring gave the spectra. Nicolson predicted the frequencies emitted by a straightforward classical analysis of these oscillation frequencies, in the spirit of earlier work on the plum pudding model.

He did bring in Planck's constant, though. He knew that dimensionally it was a unit of angular momentum, and he suggested that the atom could only lose angular momentum in discrete amountspresumably constant multiples of h. Nicolson felt that, given the dimensionality of Planck's constant, quantization of angular momentum was more plausible than quantization of energy. Of course, for the simple harmonic oscillator they amounted to the same thing, but not for any other system.

Bohr Returns to Denmark

Bohr left Manchester in July 1912 and was married on the first of August. In the fall, he began work at the University of Copenhagen. At the same time, he began setting down on paper some of his Manchester ideas about atoms. He read Nicolson's work.

As he wrote to Rutherford at the end of January 1913, he and Nicolson were really looking at different thingsNicolson was considering atoms in a very hot environment (like the sun's corona, or an electrical discharge tube) and the spectra gave information about how energy was emitted as the atom settled into its ground state. Bohr himself was only interested in the state in which the system possessed the smallest amount of energy. He went on: " The question of calculation of the frequencies corresponding to the visible part of the spectrum". At that time, Bohr thought of spectra as pretty but peripheral, having as little to do with basic physics as the colours of a butterfly had to do with basic biology.

Bohr Changes his Mind about Spectra

In February 1913, Bohr was surprised to find out in a casual conversation with the spectroscopist H. R. Hansen that some patterns had been discerned in the apparent chaos of spectral lines.

In particular, Hansen (a colleague and former classmate of Bohr) showed him Balmer's formula for hydrogen. They had very likely seen this in class together, but, given Bohr's opinion of the value of spectra, he probably hadn't paid much attention.

Balmer's formula is:

$$\frac{1}{\lambda} = R_H\left(\frac{1}{4} - \frac{1}{n^2}\right)$$

for the sequence of wavelengths of light emitted, with $n = 3, 4, 5, 6$ being in the visible, the lines used by Balmer in finding the formula. Hansen would doubtless have informed Bohr that the 1/4 could be replaced by $1/m^2$, with m another integer. The constant appearing on the right hand side is called the *Rydberg constant*, $R_H = 109{,}737$ cm^{-1}. (This is the modern value-Balmer got it right to one part in 10,000, about the limit of spectral measurements at the time.)

Bohr said later: "As soon as I saw Balmer's formula, the whole thing was immediately clear to me." What he saw was that the set of allowed *frequencies* (proportional to inverse wavelengths) emitted by the hydrogen atom could all be expressed as *differences.*

This immediately suggested to him a generalization of his idea of a "stationary state" lowest energy level, in which the electron did not radiate. There must be a *whole sequence* of these stationary states, with radiation only taking place as the atom jumps from one to another of lower energy, emitting a single quantum of frequency f such that

$$hf = E_n - E_m,$$

the difference between the energies of the two states. Evidently, from the Balmer formula and its extension to general integers m, n, these allowed non-radiating orbits, the stationary states, could be labeled 1, 2, 3,..., n,... and had energies –1, –1/4, –1/9,..., $-1/n^2$,... in units of hcR_H (using l $f = c$ and the Balmer equation above).

The energies are of course negative, because these are bound states, and we count energy zero from where the two particles are infinitely far apart. Bohr was very familiar with the dynamics of simple circular orbits in an inverse square field. He knew that if the energy of the orbit was $-hcR_H/n^2$, that meant the kinetic energy of the electron, $\frac{1}{2}mv^2 = hcR_H/n^2$, and the potential energy would be.

$$-(1/4\text{pe}_0)e^2/r = -2hcR_H/n^2.$$

It immediately follows that the *radius* of the nth orbit is proportional to n^2, and the *speed* in that orbit is proportional to $1/n$. *I*t then follows that the angular momentum of the nth orbit is just proportional to n. Evidently, then the angular momentum in the nth orbit was nKh, where h is Planck's constant and K is some multiplying factor, the same for all the orbits, still to be determined.

In fact, the value of K follows from the results above. R_H, m, h, and c are all known quantities (R_H being measured experimentally by observing the lines in the Balmer series) so the above formulas immediately give the electron's speed and distance from the nucleus in the nth orbit, and hence its angular momentum. Therefore, by putting in these experimentally determined quantities, we can find K.

Bohr Finds the Rydberg Constant without Doing an Experiment

The Balmer formula gave Bohr the essential clue that led to the realisation that the angular momentum was quantized: $L = Kh$, $2Kh$, $3Kh$,... where h is Planck's constant, as usual, and K is some constant numerical factor, presumably of order 1. Bohr gave a very clever argument to find K without doing any experiment.

First, think about how the size of K affects the *physical* properties of the hydrogen atom. How would the atom be different for $K = 10$ compared with $K = 1$? For $K = 1$, the allowed orbits would be those having angular momentum h, $2h$, $3h$, $4h$,.... For $K = 10$, the only allowed orbits would be those having angular momentum $10h$, $20h$,.... Evidently, for $K = 10$ there will be a lot fewer spectral lines, and the average *spacing* between them will be *greater* that for $K = 1$. Next, Bohr imagined a really immense hydrogen atom, an electron going around a proton in a circle of one meter radius, say. This would have to be done in the depths of space, but really this is just a thought experiment in the spirit of Einstein.

The point is that for this very large atom, the electron is moving rather slowly over a distance scale we are familiar with. We know from many experiments that charges moving at these slow speeds over ordinary (human size) distances emit radiation according to Maxwell's equations.

Or, more simply, if it's going round the circle at frequency f revolutions per second, it will be emitting radiation at that frequency $f$because its electric field, as seen from some fixed point a meter or so away, say, will be rotating f times per second.

On the other hand, the angular momentum quantization condition must be true for all circular orbits of the electron around the proton, even for this very large atom. Furthermore, the radiation emitted must still be given by the difference in energies of neighbouring orbits,

$$hf = E_{n+1} - E_n.$$

But $E_{n+1} - E_n$, the energy *spacing* between neighbouring orbits, *depends on K*. Therefore, Bohr concluded *K is fixed* by requiring that the frequency of radiation emitted by a really large atom be correctly given by ordinary common sense-that is, the frequency of the radiation must be the same as the orbital frequency of the electron, the number of cycles a second. In other words, for a large orbit we must have

$$E_{n+1} - E_n = hf = h.v/2\pi r$$

Here v is the speed of the electron in the orbit, and the orbit radius is r. *The strategy is then as follows:* We assume that the only allowed orbits are those having angular momentum integral multiples of Kh, where K is some constant, so the nth orbit has angular momentum nKh.

We can then use the equation of motion to determine the radius r_n, the electron speed v_n and the energy E_n for the nth orbit. Naturally, these all depend on K. *Now concentrate on orbits close to one meter in radius*. They will radiate at the frequency given by the equations of motion for an electron circling a proton one meter away.

But this must *match up* with the frequency given by the energy difference between neighbouring orbits divided by h. Now, if K is very small, this energy difference is small, and they won't match. If K is very large they won't match either. We must find the value of K for which these frequencies do match. That is what we do below in detail.

We establish below that there can only be agreement between the classical radiation frequency for a man-sized atom and Bohr's prediction if $K = 1/2\pi$. Therefore, we must assume the angular momentum is always quantized in chunks of size $h/2\pi$.

It follows that the allowed energy levels are

$$E_n = -\frac{1}{4\pi\varepsilon_0}\cdot\frac{e^2}{2r_n} = \left(\frac{1}{4\pi\varepsilon_0}\right)^2\cdot\frac{me^4}{2K^2h^2}\cdot\frac{1}{n^2}$$

$$= -\left(\frac{1}{4\pi\varepsilon_0}\right)^2\cdot\frac{2\pi^2 me^4}{h^2}\cdot\frac{1}{n^2}.$$

Putting this together with $E_n - E_m = hf$ we get the Balmer formula:

$$f = \frac{E_n - E_2}{h} = \left(\frac{1}{4\pi\varepsilon_0}\right)^2\cdot\frac{2\pi^2 me^4}{h^3}\cdot\left(\frac{1}{4} - \frac{1}{n^2}\right).$$

The new point is that there is no adjustable parameter! The Rydberg constant that appeared before is here given in terms of h, m and e.

The rather abstract argument that the quantum predictions must match the known classical results for large slow systems actually fixes the Rydberg constant. That is to say,

$$R_H = \left(\frac{1}{4\pi\varepsilon_0}\right)^2\cdot\frac{2\pi^2 me^4}{ch^3}$$

This formula was found to be correct within the limits of experimental error in measuring the quantities on the right. This matching for large systems is called the Correspondence Principle: in that limit, quantum predictions must correspond to known classical results.

Derivation of the Angular Momentum Quantization from the Correspondence Principle

Let us assume this large orbit is the nth (where n is of order 10^5!), so it has angular momentum $mvr = L = nKh$.

Using,

$$\frac{mv^2}{r} = \frac{1}{4\pi\varepsilon_0}.\frac{e^2}{r^2},$$

we find,

$$m\sqrt{\frac{1}{4\pi\varepsilon_0}.\frac{e^2}{mr}} \; . \; r = nKh$$

from which we find the radii of the allowed orbits are given by

$$r_n = \frac{4\pi\varepsilon_0}{me^2}.n^2K^2h^2 .$$

Therefore the allowed energies are:

$$E_n = -\frac{1}{4\pi\varepsilon_0}.\frac{e^2}{2r_n} = -\left(\frac{1}{4\pi\varepsilon_0}\right).\frac{me^4}{2K^2h^2}.\frac{1}{n^2}$$

Thus for n *very large*:

$$E_{n+1} - E_n = -\left(\frac{1}{4\pi\varepsilon_0}\right)^2.\frac{me^4}{2K^2h^2}.\left(\frac{1}{(n+1)^2} - \frac{1}{n^2}\right)$$

$$= \left(\frac{1}{4\pi\varepsilon_0}\right)^2.\frac{me^4}{2K^2h^2}.\frac{2}{n^3}$$

Now this must be equal to $hf = hv/2\pi r$ for the appropriate v, r for this large orbit.

Using $mv_nr_n = nKh$, $v_n = nKh/mr_n$.

Thus $hv/2\pi\ r = nKh^2/2\pi\ mr_n^2$.

Putting now

$$E_{n+1} - E_n = hf = h.v/2\text{p}\ r = nKh^2/2\text{p}\ mr_n^2$$

And using

$$R_n = -\left(\frac{4\pi\varepsilon_0}{me^2}\right).n^2 2K^2h^2$$

gives,

$$E_{n+1} - E_n = \left(\frac{1}{4\pi\varepsilon_0}\right)^2.\frac{me^4}{2K^2h^2}.\frac{2}{n^3}$$

$$= \frac{nKh^2}{2\pi m r_n^2} = \frac{nKh^2}{2\pi} \cdot \left(\frac{1}{4\pi\varepsilon_0}\right)^2 \cdot \frac{me^4}{n^4 K^4 h^4}$$

Now, the second and fourth terms in the above equation must be equal in the limit of large n. Canceling out common factors of the two, we find that the condition for equality is:

$$K = 1/2\pi.$$

This was the argument Bohr used to establish that angular momentum for his model is quantized in units $h/2\pi$.

Reality of Atoms

Speculative Atomism

Actual proof of the reality of atoms is an achievement of this century; even at its beginning critical consideration required the confession that atoms, although investigators spoke of them, were still no more than a stimulating and useful hypothesis, which in the final analysis was unproved and possibly unnecessary, perhaps even misleading. Although the idea of atomism was important and fruitful at the very beginning of western natural research, this idea did not originate from western research itself; just as we borrowed the model for mathematical thinking from the Greeks, we took the atomistic interpretation of nature from them.

Nothing exists — Democritus taught — except atoms and vacuum; everything else is imagination. These innumerable atoms, indestructible and invariable elementary constituents, are to be considered the basis for all beings and events in nature.

Individual atoms possess invariable geometric form and motions which fluctuate due to the pressure of and collision with other atoms. In all changes in the structure of nature the atoms are always preserved; from nothing comes nothing; nothing which exists can be destroyed; change is merely combination and separation of parts.

The differences of all things stem from the diversity of their atoms in number, magnitude, structure and arrangement. Since the movements of each individual atom are regulated according to natural law, nothing arbitrary can happen in all nature; nothing happens accidentally, but everything from a cause and of necessity.

Our rough senses can not recognize these amazingly fine elements in their true nature and form, but experience their effects only vaguely. "Only in the imagination does sweetness exist, in the imagination bitterness, in the imagination heat, cold, colours; in reality nothing exists but atoms and vacuum." This is marked evidence of that "decoloration" of the world, the assurance that all direct sensory observation is an illusion and that a knowledge of the true status of things must lead us to a picture in which nothing remains but the geometrical form and mechanical motion of the atoms. These idea forms of materialistic philosophy introduce the possibility of considering all events in nature as results of strict regularity; the gist of the basic concept of natural scientific thought is anticipated here in general and is thoroughly formed for the first time.

On the other hand we must not forget to recognize how radically this philosophy opposes religious representations of its time. For the Greeks every stream was a God; in the springs lived the nymphs, in the woods the horned Pan, and in caves and caverns resided demons.

This entire mythological picture of the world, which suspected the arbitrary and incalculable influence of demoniacal powers in every natural process, was pushed aside by the powerful conception of a picture of nature of stricter regularity. Epicurus, who renewed Democritus' teaching, and who in his thinking and conduct of life combined a rationalistic, vigourous attitude with a very cultivated, fine-spirited being, was not so brutal as to deny existence completely to the Gods.

He let them continue as blessed, immortal beings, occupied with themselves, and never infringing upon the workings of the world — these workings being developed according to the mechanical lawfulness of atomic motions in strict, definite sequence. In a great didactic poem *"De rerum natura"* the Roman Lucretius explained completely the ideas of atomistic regularity for all the fields of nature known at that time. These teachings were made known to western scholars in connection with the humanistic studies of the Renaissance; Gassendi, principally, introduced Greek philosophy into western science.

Let us consider the ideas which Newton, together with Gassendi, applying the divergent theories of Descartes, formed about the construction of matter. F. Dannemann explains them as follows in his historical work on the natural sciences: "He considered it most plausible, that it (matter) consists of solid, impermeable, moving particles.

Since natural bodies, e.g., water, are invariable in their properties, the particles of which they consist must neither be able to be used up nor destroyed. The variation of material things is to be laid exclusively to the separations, combinations and movements of those invariable particles." These are, as is obvious, exactly the concepts of materialistic philosophy, almost unchanged, just as Democritus stated them. But Newton was — outside of natural science — in no wise an adherent of materialistic theories. For him these ideas meant nothing but a clue to physical research; he did not see in them the support of radical, world-viewing results.

Quantized Atom

A very interesting fact about atoms has been known since the middle of the 19th century. When atoms are irradiated with white light (light consisting of photons of all wavelengths), they absorb only certain definite frequencies of the light, not all frequencies. Atoms of different elements absorb different frequencies. How do we interpret this? We have seen that the frequency of light is a measure of its energy. Thus atoms absorb only light of certain energies.

This implies that the atom is only allowed to have certain energies. We now know that the electric field associated with the light photon interacts with the electrons of the atom. When a photon is absorbed by an atom, the energy of the photon is used to cause an electron of the atom to make a transition ("jump") to a higher energy state within the atom.

The Absorption of Light by Atoms and Molecules

We begin this section by stating some very important ideas, which will then be illustrated in subsequent discussion. The ideas are these:

- Light consists of oscillating electric and magnetic fields.
- Because nuclei and electrons are charged particles, their motions in atoms and molecules generate oscillating electric fields.
- An atom or molecule can absorb energy from light if the frequency of the light oscillation and the frequency of the electron or molecular "transition motion" match. Unless these frequencies match, light absorption cannot occur. The "transition motion" frequency is related to the frequencies of motion in the higher and lower energy states.
- By measuring the frequencies of light absorbed by an atom or molecule, we can determine the frequencies of the various transition motions that the atom or molecule can have. Thus we can use light absorption to probe the dynamics of atoms and molecules!

These ideas are the basis for the techniques of *spectroscopy*, the study of the interaction of light and matter. The word means "measurement of spectrum," where the spectrum of a substance is the array of frequencies of light absorbed by its atoms or molecules. You should already be familiar with the idea of energy absorption through matching frequencies. This is exactly the way in which you transfer energy to a child on a swing. You push at a frequency that matches that of the swing.

Energy is transferred successfully, making the swing go to a higher energy state (higher amplitude), only if the push is made at the right time. If the frequency of pushing does not match the frequency of swinging, very little if any energy is transferred. In fact, the swing may actually lose energy if you try to push it while it is moving back towards you. This idea operates in the same way when light and matter interact. It is complicated only by the fact that, in contrast to the swing, molecules can undergo more than one type of motion.

Molecular Motions and Energy

There are many ways in which a molecule can move, and each motion contributes to the energy of the molecule. *Molecular motion is quantized* — that is, each motion can occur with only certain energies. A molecule may gain or lose energy in one of these motions by absorbing or emitting a photon of light.

The photon has energy that matches the allowed energy gain or loss associated with the motion. By measuring the energies of the photons absorbed by the molecule, we obtain its spectrum. The spectrum provides information about the motions of the molecule.

We begin with an expression for the energy of a molecule, written as a sum of the energies of the various motions possible to it. Each motion will then be described, along with the region of the electromagnetic spectrum with which it interacts. The total energy of a molecule may be written as follows:

$$E_{total} = E_{spin} + E_{translation} + E_{rotation} + E_{vibration} + E_{electrons} + E_{nucleus}$$

Nuclear Magnetic Resonance—Espin

The nuclei of some (but not all) atoms behave as if they were rotating, or spinning, about an axis passed through them, much as a top spins about its central axis. This is quite easily visualized in terms of the particle nature of the nucleus; it is less easily visualized in terms of the wave picture, which by now we know is another face of nature that is manifested in the atomic/molecular realm. Spin of nuclei is similar to electron spin.

Examples of atoms whose nuclei spin are hydrogen, fluorine, the 13 isotope of carbon, and the 15 isotope of nitrogen. The spin of the hydrogen nucleus, because hydrogen atoms occur very frequently in molecules and are therefore very useful in determination of structure. When placed in a strong magnetic field, the spinning motion of the hydrogen atom gives rise to two energy levels. The spin can be promoted from the lower to the higher level by interacting with and absorbing energy from electromagnetic radiation in the radio frequency (RF) region of the spectrum.

A typical radio photon has a frequency of 10^8 s^{-1}, a wavelength of 3 meters (about ten feet), and an energy of 6.6×10^{-26} Joules. The separation between the two spin energy levels is a sensitive function of the molecular environment of the hydrogen atom in a molecule. Thus hydrogen atoms in different environments absorb photons of different energies.

By plotting the amount of energy absorbed by the spinning nuclei versus the frequency of the RF radiation applied to the molecule, we obtain the *nuclear magnetic resonance (NMR) spectrum* of the molecule. This spectrum provides information about the chemical environment of the spinning hydrogen nucleus, and can be used to deduce the atomic bonding patterns in the molecule.

NMR has now been developed to such a level of sophistication that it can be used to determine the 3-dimensional structures of huge protein molecules in solution. Recently, nuclear magnetic resonance spectroscopy has found medical applications, in which context it is known as magnetic resonance imaging (MRI). MRI utilizes the spinning motions of atomic nuclei to provide a map of the internal structure of human body tissue.

We will now discuss the nature of NMR spectra in a systematic way, beginning with spectra of simple molecules, and gradually escalating in complexity. As we proceed through the discussion, we will introduce some of the jargon used by practicing chemists when discussing NMR. As a first bit of jargon, we will henceforth refer to hydrogen atoms as *protons*.

This is appropriate, because it is the nucleus of the hydrogen atom that is spinning, and this nucleus indeed consists of a single proton. We begin by considering the NMR spectrum resulting from the molecule. The molecule has a single proton on the right (on carbon 1), and a pair of protons on the left (on carbon 3). Carbon atoms 1 and 3 are separated by a carbon atom attached to two chlorine atoms. The single proton on the right is attached to a carbon atom that is attached to 2 chlorine atoms.

In contrast, each proton on the left is attached to a carbon atom that is attached to one chlorine atom and a second hydrogen atom. The environments of the protons on carbon 1 and carbon 3 are clearly different. The energy level spacing of the two types of protons should thus be different, and two signals, one for each type of proton, should appear in the NMR spectrum. Several aspects of this spectrum are important. First, the single sharp signal at exactly 0 ppm is not due to the subject molecule, but is instead due to a reference compound named *tetramethylsilane*, or TMS. By general agreement, the positions of NMR signals are reported relative to the position of the signal for TMS. Second, the unit used to measure resonance position, ppm (parts per million), is not a unit of energy. It has no units at all.

It may be regarded as a measure of the shift in position between the TMS resonance and a particular resonance in the compound of interest. It is usually called the *chemical shift*. The chemical shift indicates the chemical environment in which the spinning nucleus is found, as we will see shortly. Ignoring the TMS signal, then, which is present in all NMR spectra, we see that there are indeed two signals. Interestingly, the sizes, or intensities, of the signals are different. The right signal appears to be twice as tall as the one on the left. This suggests that the right signal is due to the pair of protons on carbon 3, and the left signal to the single proton on carbon 1. It also appears that there is a relationship between the position of a signal in the spectrum and the number of chlorine atoms in the immediate environment of the proton(s) giving rise to the signal. We summarize our observations of the simple spectrum in Figure in terms of three rules:

- *Rule:* Each unique proton environment generates one signal.
- *Rule:* The intensity (size) of an NMR signal gives a measure of the number of equivalent atoms responsible for the signal. A two-proton signal is twice as large as a one-proton signal.
- *Rule:* The greater the number of electronegative atoms near a proton, the further to the left of TMS the proton signal will be.

Rule 3 requires some comment. The signal due to the hydrogen atom on carbon 1 is shifted further away from TMS than is the signal of the hydrogens on carbon 3. In NMR jargon, the left signal is said to be further *downfield* (meaning further to the left) than the right signal. The reason for this is that the carbon-1 proton is close to two electronegative chlorine atoms.

These pull electrons away from carbon 1, which in turn pulls them away from the hydrogen atom. The less electron density a hydrogen atom has around it, the further downfield its NMR resonance occurs. This type of consideration helps us to assign signals to particular hydrogen atoms in molecules.

Figure shows a molecule that appears simpler than the molecule in Figure, but is in fact somewhat more complex in NMR terms. The molecule can be obtained from that in Figure by removing the central carbon atom, bringing the proton-bearing carbons together.

Clearly the carbon 1 proton is still in a different environment than are the pair of carbon 2 protons, so we might expect the NMR spectrum of this new molecule to be very similar to that of the three-carbon molecule. Specifically, we might expect two signals, one twice as intense as the other, with the less intense one downfield. The spectrum is shown in Figure.

Although there are two signals, in roughly the expected locations, the signals appear to be split into subsignals. The upfield signal now consists of two, and the downfield signal of three, closely-spaced subsignals. Chemists call the upfield signal a *doublet*, and the downfield signal a *triplet*, to indicate the number of subsignals. Apparently, there is a complication showing up here that is not present in the spectrum in Figure. As we will see, this complication is what makes NMR so useful for determination of structure.

A spinning nucleus is, in fact, a tiny magnet. When it is close enough to another proton, the second proton feels the small magnetic field of its neighbour as a tiny perturbation on the large magnetic field applied to the sample. The effect of this proton-proton interaction is to cause small changes in the energy level spacings of the proton that show up as the splittings. Interestingly, the number of subsignals into which a signal is split appears to be one greater than the number of neighbouring protons. We can now add a couple of new rules to our list:

- *Rule:* Protons on neighbouring atoms interact with each other, and split each other's signals.
- *Rule:* The number of subsignals in a *multiplet* is one more than the number of neighbour protons.

Clearly the splitting of signals due to neighbouring atoms (called *spin-spin splitting* in the jargon of the chemist) is of potentially great value in assigning signals and in determining structure; it tells us the number of protons on the atom next door to the atom bearing the protons being observed.

To test our rules, we consider the molecule in Figure. Again we see only two different proton environments here. The two carbon-1 protons are in one environment; the three carbon-2 protons are in a different environment. We expect two signals, one of total intensity two and the other of total intensity three. The signal of intensity two should be further downfield because the carbon 1 protons have an electronegative chlorine atom in their neighbourhood. We expect the downfield signal to be split to a quartet due to interaction with the three protons on carbon 2; and the upfield signal to be split to a triplet via interaction with the two protons on carbon 1. The spectrum is in Figure, and appears much as we expected it to. The NMR rules seem to work fine.

We consider one final example, the NMR spectrum of isobutyraldehyde, which will reveal a couple of other useful subtleties of NMR spectra. The molecule of interest and its NMR spectrum are shown in Figure and Figure. The protons labelled "a" in the figure experience the same environment (this is clear from a model of the molecule), and are expected to give one signal.

This signal should be the furthest upfield because these protons are furthest removed from the electonegative oxygen atom. Because there is a single proton, "b", on the neighbouring carbon atom, the signal for protons "a" should be a doublet with total relative intensity 6. This is clearly seen in the spectrum. Proton "c" should give the furthest downfield signal because it is attached to an oxygen-bearing carbon atom.

This, too, is expected to be a doublet due to interaction with proton "b". Although the splitting of the proton "c" signal is small, it is discernible in Figure. Note that the signal from "c" is very far downfield, much more so than the signals influenced by chlorine atoms in previous examples. This results from the very high electronegativity of the oxygen atom, which substantially depletes the electron density of the attached carbon atom.

Finally, we examine the signal for proton "b". This is our first encounter with a proton that interacts with two other sets of protons, and it is important to understand how the splitting works in a situation like this. Figure shows a magnified view of the signal for proton "b", which occurs at a chemical shift of 6.4 in. The signal shows a total of 14 lines, which is many more than we have encountered in a multiplet up to this point.

The explanation for all of these lines runs as follows. First, the signal for "b" is split to 7 lines by the interaction with the 6 "a"-type protons on neighbouring carbon atoms. Then, each component of the 7-line pattern is split to two lines by the interaction with proton "c". The total is 14 lines. In terms of this interpretation, Figure clearly shows a 7-line pattern, each component of which is split into two lines. This pattern is called a doublet of septets, to indicate that it results from interaction of "b" with two distinctly different types of protons. On the basis of the isobutyraldehyde example, we are in a position to state two additional rules.

- *Rule:* A proton on an oxygen-bearing carbon is shifted very far downfield.
- *Rule:* A proton with "n" neighbouring protons on one side and "m" on the other will give a signal split into (n+1)(m+1) components.

The signal should be split into 3 components from the effect of the spins of the two H atoms on the middle carbon. The signal of intensity 1, due to hydrogen "c", is furthest left because this hydrogen atom is near the oxygen atom. This signal also will be split into 3 components due to the effect of the spinning motions of the two H atoms on the middle carbon atom. Finally, the signal due to hydrogens "b" will appear at an intermediate position, with relative intensity 2. It will be split to two components by hydrogen "c"; and each of these will be split to 4 components by hydrogens "a". This signal will consist of 8 components altogether. Chemists have invented a shorthand system for presenting the information in an NMR spectrum without having to have the spectrum itself (i.e., the picture) available.

Each signal in the spectrum is described by stating, first, its multiplicity (the amount of splitting); second, the chemical shift in parts per million downfield from TMS; and third, the relative intensity of the signal, in terms of number of identical protons giving rise to the signal. The propionaldehyde spectrum is described as follows in this shorthand system:

Triplet, d = 1.1 (3H)

Quartet of doublets, d = 2.45 (2H)

Triplet, d = 9.7 (1H)

The Greek letter, d, symbolizes the chemical shift in units of parts per million. This brief description is sufficient to allow us to visualize and, if desired, even draw the spectrum, and to interpret it in terms of structure. In this section, we have considered a number of NMR spectra and from them have developed a simple set of rules for their interpretation and prediction.

It is important to realise that, although these rules are very useful in understanding the NMR spectra of relatively simple molecules, the spectra of many molecules involve complexities that our simple rules are unable to handle.

Spin-spin Splitting—A Closer Look

To this point, we have discussed the most basic aspects of nuclear magnetic resonance spectroscopy. We have acknowledged the existence of spin-spin splitting, but have not said much about its molecular origins, and have not explained the origin of the number of subsignals in a multiplet. In this section we will pursue these matters in a bit more depth. As a basis for discussion, consider a molecule in which there are three protons on carbon 1 and two protons on the neighbouring carbon atom 2.

Our rules tell us that the signal due to carbon-1 protons will be split to a triplet, and that the signal due to carbon-2 protons will be split to a quartet. Why a triplet? Why a quartet? Why do the subsignals of the triplet have relative intensities 1:2:1, while those of the quartet have relative intensities 1:3:3:1? What is the significance of the *magnitude* of the splitting between the components of a multiplet?

Focus first on the splitting of the signal due to the carbon-1 protons. The spinning protons on carbon 2 behave like tiny magnets. The magnetic field is generated by the spinning negative charge, which is much like an electric current. The direction of the magnetic field generated by a carbon 2 proton depends on its direction of spin (clockwise or counterclockwise). Either direction is equally probable for each of the three carbon-2 protons. Given this, it is easy to appreciate that there are 3 possible combinations of spin directions for the 2 protons on carbon 2.

Both can be spinning clockwise, in which case their magnetic fields add together and reinforce the large laboratory magnetic field. One can be spinning clockwise and the other counterclockwise, in which case their magnetic fields cancel. This arrangement can occur in two different ways because either proton 1 or proton 2 can have clockwise spin. The arrangement of one spin "up" (cw) and the other "down" (ccw) is thus twice as probable as the arrangement with both spins "up".

Finally, both protons can have spin "down", in which case their magnetic fields add together but oppose the applied laboratory field. The spin arrangements of the carbon-2 protons are perhaps better appreciated when presented pictorially. We conclude that the carbon-1 protons will feel three slightly different magnetic fields, one slightly larger than, one equal to, and one slightly smaller than the applied field. The middle possibility is twice as probable as the other two. Now we are in a position to appreciate why the signal due to the carbon-1 protons is split to 3 subsignals of relative intensities 1:2:1. This is a simple consequence of the fact that the carbon-2 protons create 3 slightly different magnetic fields that can be experienced by the carbon-1 protons. One signal arises from each of these magnetic fields, with intensity proportional to the probability of the spin arrangement giving rise to the magnetic field perturbation caused by protons on carbon 2.

In similar fashion, we can understand why the signal due to carbon-2 protons is split to 4 subsignals. The 3 protons on carbon 1 can have their spins aligned in 4 different ways (all "up" (one possibility), two "up" one "down" (3 possibilities), one "up" two "down" (3 possibilities), and all "down" (again only one possibility). The carbon-2 protons will experience 4 slightly different magnetic fields with probabilities 1:3:3:1, so will be split to a quartet whose components have intensities reflecting these probabilities. The number of components in a multiplet and their relative intensities are thus seen to have a physically plausible explanation.

In general, as we have seen, "n" neighbouring protons cause a splitting into "n+1" components. For reasons that we need not pursue, the relative intensities of the components resulting from splitting by any number of equivalent neighbouring protons can be readily computed from Pascal's Triangle, in which a particular number is calculated as the sum of the numbers to the left and right of it in the row above.

We conclude our examination of spin-spin splitting by considering the magnitude of the separation between two adjacent signals within a multiplet. We can base the initial part of the discussion on the same molecule used above, in which there are three protons (call them protons A) on carbon 1 and two protons (protons B) on the neighbouring carbon atom 2. The magnetic interaction between protons A and B is called spin-spin coupling.

Protons A are said to couple with (meaning that they interact with) protons B. The strength of the interaction determines the magnitude of the splitting between adjacent components of a multiplet. The stronger the interaction, the larger the splitting. Further, spin-spin coupling between the two sets of protons A and B obeys a version of Newton's third law, that for every action there is an equal and opposite reaction.

If protons A interact with protons B with a certain strength, splitting the B signal by a certain amount, then protons B interact with protons A with exactly the same strength, splitting the A signal by exactly the same amount. The size of the splitting within a multiplet is called the *coupling constant, J*, and is usually measured in frequency units. We will not be concerned with measuring coupling constants, so do not worry at this point about the units. The importance of coupling constants for us will be this: the signals for two sets of interacting (coupled) protons show exactly the same coupling constant. Thus it is often possible by looking at the NMR spectrum to decide which signals in the spectrum are coupled. This is an additional aid in building the structure of a molecule from an analysis of its NMR spectrum. To show how this works, we complicate our test molecule somewhat by adding a third carbon to which is attached a single proton, D.

We expect the following coupling interactions within this structure:

- A with B only; the signal for A should be a 1:2:1 triplet
- D with B only; the signal for D should be a 1:2:1 triplet
- B with A and also with D; the signal for B should be a 1:3:3:1 quartet as a result of coupling with A. Each component of the quartet should be split to a doublet as a result of coupling with D.

  Suppose that the interaction between A and B is stronger than that between B and D. For discussion sake, we will assume that B and D couple with only 1/3 the strength of A and B. Thus if the coupling constant for interaction between A and B is 12 s^{-1}, then that for interaction between B and D is 4 s^{-1}. Then we can make the following additional statements about the appearance of the NMR spectrum:
- The splitting within the multiplet due to protons A (12 s^{-1}) will be 3 times bigger than the splitting within the multiplet due to protons D.
- The B multiplet will consist of 4 pairs of signals. The members of a pair will be separated by 4 s^{-1}; the corresponding members of two adjacent pairs will be separated by 12 s^{-1}.

The spectrum of our test molecule might look as shown, where we have assumed chemical shifts of 1 ppm, 2 ppm, and 4 ppm for the A, B, and D protons, respectively.

Infrared Spectroscopy—Evibration

Infrared spectroscopy allows us to examine the vibrational motions of molecules. We will discuss these motions as one of several types of motion that the molecule as a whole can undergo.

Etranslation, Erotation, Evibration

The nuclear spinning motion discussed above involves only one atom of a molecule. The three motions to be discussed now — translation, rotation, and vibration — are motions of the molecule as a whole in space, involving concerted movements of all of the atoms. For illustration, we use a very simple molecule consisting of only 2 atoms. Such molecules are called *diatomic molecules.* Examples are O_2, H_2, HCl, and CO. A generic diatomic molecule is pictured in Figure.

First, the entire molecule may move through space in an arbitrary direction and with a particular velocity. This is called *translational motion* and with it we associate the translational kinetic energy of the molecule:

$E_{trans} = 1/2\ mv^2$ (m = mass of molecule; v = velocity of centre of mass of the molecule) The translational velocity of the molecule has components along each axis of a Cartesian coordinate system, so $E_{trans} = 1/2\ mv^2 = 1/2\ mv_x^2 + 1/2\ mv_y^2 + 1/2\ mv_z^2$, where v_x is the x-component of velocity, etc., and m is the mass of the molecule

The total translational KE of the molecule is made up of three parts, each representing the kinetic energy of the molecule along one reference direction. The molecule has 3 translational degrees of freedom (i.e., independent directions in which it can translate), one corresponding to each Cartesian axis. A molecule travelling through space at a temperature of 300 K has translational kinetic energy of 6.2×10^{-21} Joules. However, the spacing between the allowed translational energies for an HCl molecule in a container 20 cm in length is on the order of only 10^{-40} J.

This is such a small energy that transitions between translational kinetic energy levels are difficult to observe. At the present time, no spectroscopy is done involving these energy levels. Second, the molecule may rotate about an internal axis. For example, rotation of a diatomic molecule is diagrammed in Figure. Like translation, a molecular rotation may be resolved into three mutually perpendicular components. The rotational kinetic energy of the molecule is

$$KE_{rot} = 1/2\ I_x w_x^2 + 1/2\ I_y w_y^2 + 1/2\ I_z w_z^2,$$

where I_x, I_y, and I_z are the moments of inertia (the moment of inertia is the rotational equivalent of mass) about the x, y, and z axes, and w_x, w_y, w_z are the angular velocities (in units of radians/s) about these axes. There are 3 *rotational degrees of freedom*, one for each Cartesian axis. An exception arises in the case of a linear molecule, for which one of the three axes is coincident with the molecular axis. The molecule has no rotational energy about this axis, because the corresponding moment of inertia is zero (i.e., no mass lies off the axis). Linear molecules have only two rotational degrees of freedom.

Rotation is a type of oscillation, in which the number of rotations per second is the frequency. If the molecule has associated with it a permanent electric field, then rotation of the molecule about some axis causes an oscillation in the electric field. If this oscillation matches in frequency the oscillation in the electric field of an impinging photon, the molecule can absorb the photon and go to a higher energy rotational state. Rotations of molecules at room temperature typically occur with energy on the order of 10^{-23} J.

Photons with frequencies near 10^{11} s^{-1} (wavelength, 0.03 m), in the microwave region of the electromagnetic spectrum, are required to interact with such motions. The rotational (or microwave) spectrum of a molecule gives information about the moments of inertia, from which can be extracted the bond lengths and bond angles in the molecule. Interpretation of rotational spectra is complex, and we will not be concerned with it at this time.

Finally, the molecule may vibrate. Chemically bonded atoms are like masses connected by a spring (the bond). The bonds can be stretched and compressed somewhat, and the bond angles can vary within certain limits. Motions of the molecule in which bonds are compressed and/or stretched, or in which bond angles get larger then smaller, but in which the centre of mass of the molecule does not move through space, are called *vibrational motions*. A diatomic molecule vibrates by repeated stretching and contraction of the bond joining the two atoms. The molecule has one *vibrational degree of freedom* in addition to 3 translational and 2 rotational degrees of freedom.

For a polyatomic molecule the number of vibrational degrees of freedom (called *vibrational modes*) is the total number of degrees of freedom minus those of translation and rotation. The total is 3N, where N = the number of atoms in the molecule.

This is true because each atom may independently move in any of three directions and therefore has 3 degrees of freedom available to it; the molecule has a total of 3N. The number of vibrational degrees of freedom is 3N-6 for a non-linear polyatomic molecule, and 3N-5 for a linear polyatomic molecule. The water molecule, with N = 3, has $3 \times 3 - 6 = 3$ vibrational degrees of freedom.

Vibrational motions occur with characteristic frequencies (just as does the motion of the playground swing). In addition, the vibration causes a change in the distribution of the electrons in the molecule, which may give rise to an oscillating electric field. If the frequency of impinging light matches the frequency of the vibrational motion, the molecule may absorb light photons (energy) and move to a vibrational state in which the amplitude of vibration is increased.

By measuring the frequencies of light absorbed, we may observe how rapidly the molecular vibrations occur! Since the frequencies of vibrations are directly related to the strengths of the chemical bonds (i.e., to the stiffness of the spring), vibrational frequencies give us information about chemical bond strengths.

Typically, molecules vibrate at between 10^{13} and 10^{14} times per second. Photons with frequency 10^{14} s^{-1} (wavelength 3×10^{-6} m) have energy 6.6×10^{-20} J. Vibrational motions therefore have frequencies that are about 10^3 times larger than the frequencies of molecular rotation and a million times larger than nuclear spin frequencies. Higher energy photons in the infrared region of the spectrum are required to excite the vibrations.

Infrared (IR) spectroscopy deals with transitions between vibrational energy levels in molecules; it is therefore also called *vibrational spectroscopy*. Because IR spectra are easily acquired and provide much structural information, we will discuss them in some detail. An IR spectrum is a plot of the energy of the infrared radiation (expressed either in micrometers (1 mm = 10^{-6} m) or wavenumbers, the reciprocal of wavelength in centimeters) versus the per cent of light transmitted by the compound.

Within the energy range between 200 and 4000 cm^{-1}, the vibrational spectrum of the molecule appears as a series of absorption bands of variable intensity. The absorption bands are the downward-facing spikes, or peaks. Note that when absorption occurs, the per cent of light transmitted by the sample decreases from near 100 to various smaller values. Each absorption band in the spectrum corresponds to a vibrational transition within the molecule, and gives a measure of the frequency at which the vibration occurs. For the water molecule, with three vibrational degrees of freedom, there are three sets of energy levels within which transitions may occur.

The spacing between energy levels depends upon the nature of the vibration. Each spacing requires a photon of different energy to cause the transition, so infrared photons of three different energies are absorbed by H_2O. The frequencies of the photons (and of the vibrations) are 3600, 3500, and 1650 cm^{-1}, respectively, for the asymmetric stretch, the symmetric stretch, and the bend.

All molecules vibrate, but not all vibrations interact with electromagnetic radiation. For a vibrational motion to absorb infrared electromagnetic radiation, it must produce a change in the dipole moment of the molecule. HCl, for example, with a centre of positive charge located near the H atom and a centre of negative charge located near the Cl atom, has a dipole moment. Moreover, the magnitude of the dipole moment changes as the bond stretches, so this vibration is able to absorb IR radiation and is said to be *IR active*. HCl thus exhibits an IR spectrum. The N_2 molecule, on the other hand, has no dipole moment. Furthermore, stretching the NN triple bond does not change the dipole moment.

Consequently, the molecular vibration is infrared inactive (cannot directly absorb IR radiation). There are many molecules that, although possessing no permanent dipole moment, still undergo vibrations that change the value of the dipole moment from 0 to some non-zero value. Consider the CO_2 molecule. This molecule has no permanent dipole moment, since the individual bond dipoles exactly cancel. However, when it undergoes a bending vibration its dipole moment changes from zero to some non-zero value.

This vibration produces a change in dipole moment and is therefore IR active. All three vibrations of the water molecule change the dipole moment and are IR active. Three bands appear in the IR spectrum. The requirement that a vibration must cause a change in the dipole moment in order for a molecule to absorb radiation is another situation in which an oscillating electric field in the molecule must match that in the radiation. Oscillation of the dipole moment generates such an oscillating electric field.

In addition to the *number* of distinct vibrations expected for a molecule, we can anticipate their *locations* in the spectrum. For a diatomic molecule, A-B, the wavenumber, n, of the infrared radiation absorbed by vibration of the molecule is given by,

$$n_{AB} = n/c = 1/l = (1/2p)\ (k_{AB}/m_{AB})^{1/2}w$$

Here,

$$m_{AB} = M_A M_B/(M_A + M_B)$$

k_{AB} is the force constant for the A-B bond and measures the bond strength; m_{AB} is called the reduced mass; and M_A and M_B are the masses of the atoms. This equation indicates that the heavier the atoms involved in the bond, the lower the absorption frequency, given a constant bond strength. Care is required in applying this equation, however. For the series HF, HCl, HBr, and HI, the IR absorptions occur at, respectively, 3958, 2885, 2559, and 2230 cm^{-1}. It is tempting to conclude that this is due to the increase in mass of the halogen atom. Calculation of m for each molecule reveals that the reduced mass actually changes very little. The change in frequency is instead due to the decrease in bond strengh along the series. Equation suggests that, because the frequency of a vibration depends on the identities of the two bonded atoms and the strength of the bond between them, a particular type of bond should always absorb at more or less the same location in the infrared spectrum. This turns out to be true.

Thus, for example, all molecules containing C-H bonds show absorptions in the region of 3000 cm^{-1}, corresponding to the stretching motions of these bonds. To develop a feel for the manner in which structural changes in the framework of a molecule are reflected in the infrared spectrum, we will consider the IR spectra of a series of relatively simple molecules based on a chain of 5 carbon atoms. The first member of the series has formula C_5H_{12}, and is called pentane. The infrared spectrum of pentane is in Figure.

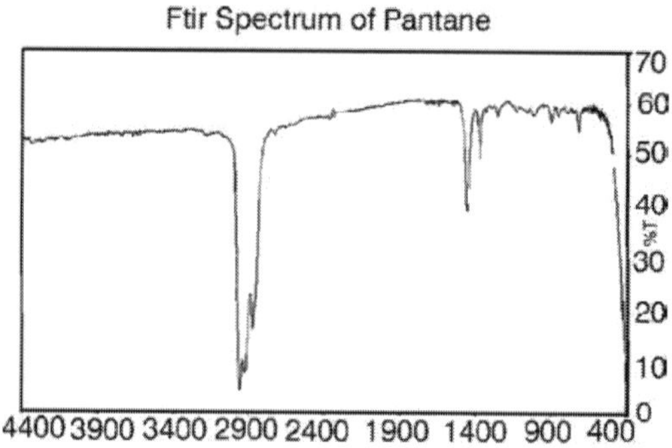

Figure: *IR Spectrum of Pentane*

The spectrum is relatively uncomplicated. There is a group of absorption bands in the region 2800-3000 cm^{-1} that correspond to C-H stretching vibrations; a pair of bands in the 1350-1500 cm^{-1} region, due to C-H bending vibrations; and a series of very weak bands in the region between 700 and 1400 cm^{-1}, due to various complex motions of the molecule that involve C-C stretches, C-C bends, and more complicated C-H motions. The most intense absorption in the spectrum is clearly that due to C-H stretching, because the largest dipole moment change arises from these motions. The IR spectrum of 1-chloropentane, $C_5H_{11}Cl$. The number preceding the name of this compound indicates that the chlorine atom is attached to the first carbon atom in the chain. The spectrum is remarkably like that of pentane, with two obvious changes.

First, the bands in the 700-1400 region, although they show the same pattern as in this region of the pentane spectrum, are more intense than those for pentane. This is a result of the chlorine atom, which makes the molecule polar, resulting in a larger change in dipole moment during many of the vibrations. Thus intensity of absorption is increased. Second, there is a strong band at 650 cm^{-1} that is absent in the pentane spectrum. This band must be associated with the C-Cl stretching motion.

Next, we consider the spectrum of 1-pentanol, $C_5H_{11}OH$. Again, the bands in the 2800-3000 and 1350-1500 regions appear much as in pentane; and those in the 700-1400 region again show a similar pattern, but with enhanced intensity due to the presence of the electronegative OH group, and with a strong band near 1050. The new features of the spectrum are the intense, broad band at 3300-3400 cm^{-1}, and the very broad feature in the 600-700 cm^{-1} region.

The high energy 3300-3400 band corresponds to stretching of the O-H bond. A band in this region is characteristic of molecules containing -OH. The broad low energy feature is also a result of the presence of OH. The band is extremely broad because the O-H group is involved in *hydrogen bonding* with O-H groups of neighbouring molecules. In most cases, molecules containing O-H groups are involved in hydrogen bonding to some extent, and produce a broad O-H stretching band in the IR spectrum. Hydrogen bonding, one of several types of *intermolecular force.*

Finally, we examine the spectrum for the pentane derivative, CH_3-C(=O)-CH_2-CH_2-CH_3, which is called 2-pentanone (the -one ending tells us that the molecule contains the C=O unit). The striking feature of this spectrum is the intense band at 1700 cm^{-1}, corresponding to the stretching vibration of the C=O group. All molecules containing the C=O group have a band in the region 1680-1750 cm^{-1} in the infrared spectrum. The presence of such a band is an indication of this structural feature.

This brief discussion of the IR spectra of pentane derivatives should give you an appreciation of the relationship between structure and the positions of absorption bands in the IR spectrum. Let's look at two further examples.

Example. The structure and infrared spectrum of acetone, C_3H_6O. What structural information can be determined from this spectrum?

Solution. The bands just above 2800 cm^{-1} are due to C-H stretching motions, which occur at high frequency because the H atom has such small mass. Any compound containing C-H bonds will have an absorption band in this region. The intense absorption band at 1730 cm^{-1} is due to the stretch of the C=O bond. C=O bonds invariably absorb in the region 1750-1680 cm^{-1}, independent of the other groups bonded to carbon.

The appearance of a band in this region in the spectrum of an unknown compound is diagnostic of a carbon-oxygen double bond. The bands at energy lower than 1500 cm^{-1} arise from bending motions of C-H bonds and stretching and bending motions of C-C bonds. As a general rule, it is difficult to assign bands in this region to particular bonds in the molecule.

Example. The structure and infrared spectrum of t-butanol, $(CH_3)_3COH$. What structural information is available from the spectrum?

Solution. The most prominent feature of the spectrum is the strong, broad band at about 3400 cm^{-1}. This is due to the stretching vibration of the O-H bond. The band is strong (i.e. intense) because there is a substantial change in dipole moment when the O-H bond stretches. It is broad as a result of hydrogen bonding, which is the interaction between the hydrogen atom of the O-H group of one molecule with the oxygen atom of the O-H group of another. Hydrogen bonding, an example of an intermolecular force. Many molecules containing an OH group exhibit a similar vibrational band in the region 3200-3600 cm^{-1}. Next, the intense series of bands between 2800 and 3000 cm^{-1} is due to stretching motions of the various C-H bonds in the molecule. That they all occur at 3000 or less indicates that the hydrogen atoms are attached to carbon atoms that form single bonds only (rather than, say, double bonds).

Finally, the intense, somewhat broad band at 1200 cm^{-1} is due to the stretching vibration of the C-O single bond. The remaining bands cannot be assigned to particular bond types; instead, they are due to complex, interdependent vibrational motions of the molecule. However, the pattern of bands in the region below 1400 cm^{-1}, usually called the *fingerprint region*, is unique to a particular molecule, and can be used to identify it.

Functional Group Analysis and Fingerprinting

An *organic functional group* is a small group of atoms, bonded together in a characteristic way, that occurs in a large variety of compounds. Examples are C=O (carbonyl), C(=O)OH (carboxyl), and NH_2 (amine).

The functional group is often emphasized in writing the formula of a compound containing it. Thus R-COOH represents any molecule containing the carboxyl functional group. R is a generic symbol for an organic "radical", consisting of primarily carbon and hydrogen atoms bonded together, and in which one carbon atom is capable of forming one additional bond.

Organic functional groups differ from one another both in the strength of the bond(s) and in the masses of the atoms involved. For instance, the O-H and C=O functional groups contain atoms of different masses connected by bonds of different strengths.

According to equation, we therefore expect the O-H and C=O groups to absorb IR radiation at different positions in the spectrum. The presence of a strong, broad band between 3200 and 3400 cm^{-1} indicates the presence of an O-H group in the molecule, while the presence of a strong band around 1700 cm^{-1} confirms the presence of a C=O group. For organic molecules, the infrared spectrum can be divided into three regions. Absorptions between 4000 and 1300 cm^{-1} are primarily due to specific functional groups and bond types. Those between 1300 and 900 cm^{-1}, the *fingerprint region*, arise from more complex vibrations in the molecules; and those between 900 and 650 cm^{-1} indicate benzene rings in the molecule.

We will briefly discuss the characteristic infrared absorption frequencies of a variety of common organic functional groups. While studying these, realise that the infrared spectrum of even a simple compound can be very complex. Some absorption bands are weak while others are strong. Some are close enough together to coincide or overlap. Interactions such as hydrogen bonding cause absorption bands to shift and broaden.

Table: *Important Functional Groups*

Functional Group	*Wavenumber Range*
O-H	3400-3200
N-H	3300-3000
C(3)C-H	3300
=C-H	3100-3000
C(=O)-H	2700
C=O	1700
C=C	1680-1600
Benzene	1600
C-C-O	1260-1000
strong bands characteristic of benzene	900-650

Accordingly, do not expect to be able to identify every absorption band in a spectrum. Rather, look for the presence or absence of absorption bands in their characteristic spectral regions to confirm the presence or absence of particular functional groups.

Alkanes: Alkanes are hydrocarbons containing only single bonds. The most prominent bands in infrared spectra of alkanes and from alkane portions of more complicated organic compounds are due to C-H stretching and bending. The symmetric and asymmetric C-H stretching frequencies are in the regions 3000 to 2900 and 2900 to 2800 cm^{-1} respectively, and are usually weak. Because most organic compounds contain several $-CH_3$, CH_2, and/or -CH groupings, there is often overlap of absorption bands in this region.

Thus the presence or absence of bands is taken to indicate the presence or absence of C-H bonds in the molecule.

The major bending modes of CH_2 and CH_3 groups appear at 1470-1420 and 1380-1340 cm^{-1}. The exact positions depend upon the nature of adjacent atoms. Interpretation is usually complicated by the presence of several bands of this type, or additional bands from other sources. The (usually) large number of C-C bonds in an organic molecule makes the C-C stretching vibrations in the 1300-1100 cm^{-1} region uninterpretable in most cases.

Alkenes. Alkenes are hydrocarbons containing at least one C=C double bond. When H is attached to a double-bonded C, C-H stretching bands generally appear in the region 3100-3000 cm^{-1}. Alkanes and alkenes can thus be differentiated. The C=C stretching frequency occurs in the 1675-1600 cm^{-1} region and varies with the atoms attached to the double-bonded carbons.

Table: *Alkene Bending Absorptions*

Alkene	*Wavenumber range, cm^{-1}*
$R-CH=CH_2$	1000-960 940-900
$R_2C=CH_2$	915-870
trans-RCH=CHR	990-940
cis-RCH=CHR	790-650
$R_2C=CHR$	850-790

Alkynes (or *acetylenes*) Alkynes are hydrocarbons containing a carbon-carbon triple bond, C(3)C). The C-H stretching vibration of *terminal acetylenes* (meaning that the last carbon in a chain is triple bonded to its neighbouring carbon) generally appears at 3300 cm^{-1} as a strong sharp band. The CC stretching band is found at 2150-2100 cm^{-1} if the alkyne is monosubstituted (i.e., R-C(3)C-H) and at 2270-2150

cm^{-1} if disubstituted (i.e., R-C(3)C-R'). These absorptions are usually weak. Aromatic (benzene) Rings. Aromatic C-H stretching bsorption appears in the region 3100-3000 cm^{-1} and is readily distinguished from alkane, alkene, or alkyne C-H, allowing a reliable determination of the types of carbon-bound hydrogen in the molecule.

The positions of aromatic C-H bending bands at 900-690 cm^{-1} depend upon the substitution pattern of the benzene ring as indicated in Table. In the absence of other interfering absorptions, these strong, usually sharp bands can be used to distinguish postitional isomers of substituted benzenes. Sharp bands at ~1600 and ~1500 cm^{-1} are characteristic of all benzene compounds; a band at 1580 cm^{-1} appears when the ring is *conjugated* with a substituent. (The word conjugated means the substituent has a double bond one bond removed from the benzene ring).

Table: *Absorptions of Benzene Rings*

Substitution	*Wavenumber range (cm^{-1})*
mono	755-730 710-690
1,2-di	765-730 710-690
1,3-di	800-750 710-690
1,4-di	840-800
1,2,3-tri	800-760 740-700
1,2,4-tri	880-860 820-800

Alcohols. Alcohols are compounds containing the O-H functional group. The characteristic infrared band due to the O-H stretching vibration appears at 3650-3600 cm^{-1} in dilute solution. With increasing concentration in solution, or in pure liquids or solids, intermolecular hydrogen bonding broadens the band and shifts its position to lower wavenumber (3500-3200 cm^{-1}). The infrared spectrum of pure t-butyl alcohol shows the broad O-H stretching band at 3440 cm^{-1}. In addition to the O-H stretching vibration, alcohols exhibit O-H bending and C-O stretching transitions at 1500-1300 cm^{-1} and 1220-1000 cm^{-1}, respectively. Although C-O stretching bands occur in a spectral region where there are usually many other bands, they are easy to identify because they are intense (strong). They are often *coupled to* (i.e., move in concert with) C-C absorption bands and therefore exhibit splitting.

The exact location of the C-O bands depends on the degree of branching of the carbon atom that is attached to oxygen. Thus, as Table reveals, the position of the C-O absorption band can allow distinction between *primary*, *secondary*, and *tertiary* alcohols.

In a primary alcohol, the OH group is attached to a carbon that is in turn attached to only one other carbon atom. In secondary and tertiary alcohols, the OH group is located on a carbon attached to 2 and 3 other carbons, respectively.

Table: C-O Stretching Vibration of Alcohols

Primary alcohols	1050 cm^{-1}
Secondary alcohols	1125 cm^{-1}
Tertiary alcohols	1200 cm^{-1}

Ethers. Ethers are compounds containing the C-O-C functional group. The asymmetric C-O stretching absorption of ethers appears in the region 1280-1050 cm^{-1}. As in alcohols, the position of this strong band is dependent on the nature of the attached groups.

Aldehydes and Ketones. The carbonyl, (C=O) stretching absorption is particularly important because it is present in a variety of functional groups including aldehydes, ketones, and carboxylic acids and their derivatives. Table shows the basic structures of compounds containing C=O.

Table: *Carbonyl Containing Compounds*

Name	*Structure*
Aldehyde	R-C(= O) – H
Ketone	R-C(=O) – R
Carboxylic Acid	R-C(=O) – OH
Anhydride	R-C(=O)–O –C(=O) – R
Ester	R-C(=O)–OR

The C = O stretching frequencies of aldehydes in which R is an alkane are observed at 1735-1710 cm^{-1}. Although the position of the carbonyl absorption does not allow distinction between aldehydes and ketones, the former are recognizable by the C-H stretching vibration that appears as two bands in the 2850-2700 cm^{-1} region.

Double bonds in conjugation with the carbonyl group lower the wavenumber by 25 to 50 cm^{-1}. Thus, aromatic aldehydes (benzene-C(=O)H) generally absorb at 1700-1690 cm^{-1} and diaromatic ketones at 1670-1660 cm^{-1}. Intramolecular hydrogen bonding to the carbonyl oxygen also lowers the wavenumber by 25 to 50 cm^{-1}.

Carboxylic Acids: Carboxylic acids contain the C(=O)OH functional group. The carbonyl group of acids in which R is not aromatic appears at 1730-1700 cm^{-1} and is shifted to 1720-1680 cm^{-1} when the C=O double bond is conjugated with a C=C double bond.

The most characteristic absorption of carboxylic acids is a broad band from 3300-2500 cm^{-1} due to hydrogen bonded O-H stretching. The C-H stretching vibrations appear as small bands on top of this band.

Amines: Amines are compounds containing the N atom singly bonded to R or H. Primary amines have formula NRH_2, while secondary and tertiary amines have formula NR_2H and NR_3 respectively.

Primary and secondary amines show N-H stretching vibrations in the 3500-3300 cm^{-1} region. Primary amines generally have two bands approximately 70 cm^{-1} apart due to asymmetric and symmetric stretching modes. Secondary amines show only one band.

Inter- or intramolecular hydrogen bonding broadens the absorptions and lowers the frequency. In general the intensities of N-H bands are less than of O-H bands.

The N-H bending and C-N stretching absorptions are not as strong as the corresponding alcohol bands and occur at approximately 100 cm^{-1} higher frequencies. NH_2 groups give an additional broad band at 900-700 cm^{-1} due to bending.

Nitriles: Nitriles are compounds containing the C(3)N group. A sharp, unusually strong absorption at 2260-2220 cm^{-1} due to CN stretching is characteristic of nitriles. To complete the section on IR spectroscopy, we make one broad and extremely useful generalization. Every compound, organic or inorganic, has a unique infrared spectrum.

This spectrum serves as a fingerprint for the compound. Thus if a compound prepared in the laboratory is thought to be a particular known substance, the fastest way to determine whether or not it is is to obtain the IR spectra of the compound and the known substance. If the spectra are identical (band-for-band match), then so are the substances.

Electron Motion in Molecules—Eelectrons

The electrons can change energy states within the molecule. When they do this, they change location with respect to the atomic nuclei. Motions of electrons within the molecule are associated with frequencies that are typically in the range 10^{14} to 10^{15} s^{-1}. When this frequency matches the frequency of light impinging on the molecule, the light can be absorbed, and an electron moves to a higher energy state.

The frequencies of light absorbed give us information about the motions of the electrons in the molecule. Photons with frequency 10^{15} s^{-1} have energy of 6.6×10^{-19} J, and fall in the visible and ultraviolet regions of the electromagnetic spectrum.

Spectroscopy involving changes in electron motions is therefore commonly called UV-Visible spectroscopy. It is also called electronic absorption *spectroscopy* (EAS). The wavelengths of light absorbed and the extent of absorption by a substance can be measured using an instrument called an electronic absorption spectrometer. A typical spectrometer covers the wavelength range between 1100 and 190 nm and thus ranges from the near IR region (1100-750 nm) to the near UV region (400-190 nm) ("near" means bordering on the visible region).

To measure the UV-Vis spectrum of a substance requires the preparation of a solution of the substance in an approprite solvent. The spectrum is a plot produced by the spectrometer of the amount of light absorbed (the absorbance) as a function of the wavelength of the light. The spectrum of potassium permanganate, $KMnO_4$, dissolved in water.

Note that the absoption does not consist of sharp lines, as in NMR, or even as relatively narrow signals, as in IR. Instead, absorption usually occurs over a range of wavelengths. Each absorption region is therefore called an *absorption band*, to indicate that it covers a band, or range, of wavelengths, rather than occurring at just a single wavelength.

The amount of light absorbed by the substance — the absorbance — at a particular wavelength is directly proportional to 1) the length of solution through which the light beam passes; and 2) the concentration of substance (number of molecules or moles per unit volume) in the solution.

$$A = e\ l\ M$$

where A = absorbance (amount of light absorbed); l = the length of solution through which the light passes, in cm; M = the molarity (number of moles of substance per liter) of the solution; and e = the proportionality constant relating A to l and M.

E is called the molar absorptivity of the substance at the particular wavelength at which A is measured. The most important aspect of this equation for us at this point is this: the amount of light absorbed is directly proportional to the amount of substance per unit volume of solution. This is called Beer's Law.

The mlar absorptivity is characteristic of the substance and the wavelength. Values of e for a substance are usually quoted for the wavelengths of maximum absorbance, l_{max} — that is, the tops of the absorption bands. The value of e gives a measure of the effectiveness of a molecule of substance at capturing a photon of light.

Values can range from as low as 0.01 to as high as 10^5 L/mole-cm. Note that substitution of 1000 cm^3 for 1 L educes the units of e to cm^2/mole. This shows that e measures the effective cross-sectional area of the substance with respect to capturing light. Since the e values of a substance are characteristic of it, we may use the amount of light absorbed to measure the concentration of the substance in solution.

Example. The spectrum of a mixture of $KMnO_4$ and $K_2Cr_2O_7$ dissolved in water has A = 0.266 at 565 nm and A = 0.790 at 350 nm.

For $KMnO_4$, e = 1.13×10^3 at l = 350 nm and 1.27×10^3 at 565 nm. For $K_2Cr_2O_7$, e = 2.84×10^3 at l = 350 nm and essentially zero at 565 nm. Calculate the concentrations of $KMnO_4$ and $K_2Cr_2O_7$ in the solution.

Solution. Since dichromate does not absorb at 565 nm, the absorbance at 565 is due entirely to permanganate. It's concentration can be calculated from Beer's Law:

$$[MnO_4^-] = \text{A at 565/e at 565} = 0.266/1.27\times10^3$$
$$= 2.09 \times 10^{-4}\ \text{M}$$

The absorbance at 350 nm is a sum of contributions from the two species; however, the permanganate contribution can be calculated and subtracted, allowing the dichromate concentration to be obtained:

$$A_{MnO4-}\ \text{at } 350 = (\text{e at } 350)\times[MnO_4^-]$$
$$= 1.13\times10^3\times2.09\times10^{-4}$$
$$= 0.236$$

$$A_{Cr2O72-}\ \text{at } 350 = A_{total} - A_{MnO4-}$$
$$= 0.790 - 0.236 = 0.554$$

$$[Cr_2O_7^{2-}] = (\text{A at } 350)/(\text{e at } 350)$$
$$= 0.554/2.84\times10^3$$
$$= 1.95\times10^{-4}\ \text{M}$$

Example. The complex ion, $FeL(CH_3CN)(CO)^{2+}$ has the structure shown in Figure. L represents the cyclic structure consisting of carbon, hydrogen, and nitrogen atoms.The complex decomposes when dissolved in acetonitrile, CH_3CN, according to the following equation

$$FeL(CH_3CN)(CO)^{2+} + CH_3CN \rightarrow FeL(CH_3CN)_2^{2+} + CO(g)$$

The course of this reaction can be monitored by measuring the absorbance of the solution at 556 nm, a wavelength where $FeL(CH_3CN)_2^{2+}$ absorbs but $FeL(CH_3CN)\ (CO)^{2+}$ does not. The following absorbance data are obtained at various times after the reaction begins:

Time, minutes	*Absorbance*
0	0
3	.121
8.5	.243
11.5	.302
14.5	.369
17	.397
20	.447
24	.487
28	.528
36.5	.579
45	.615
59.5	.646

The initial concentration of $FeL(CH_3CN)(CO)$ was 0.694×10^{-4} M. The value of e for $FeL(CH_3CN)_2^{2+}$ at 556 nm is 9650 $M^{-1}cm^{-1}$.

- Plot $[FeL(CH_3CN)(CO)^{2+}]$ vs time.
- What percentage of the reactant remains after 45 minutes?
- What is the average rate of the reaction over the first 15 minutes?

Solution. The concentration of $FeL(CH_3CN)(CO)^{2+}$ at each time must be calculated from the measured absorbance, the molar absorptivity of the product, $FeL(CH_3CN)_2^{2+}$, and the known initial concentration of $FeL(CH_3CN)(CO)^{2+}$. For example, at t = 3 minutes, the measured absorbance is 0.121. The corresponding $[FeL(CH_3CN)_2^{2+}]$ is

$$[FeL(CH_3CN)_2^{2+}] = A/e = 0.121/9650 = 1.25\times10^{-5}\text{ M}.$$

The reaction stoichiometry indicates that 1 mole of product is obtained from 1 mole of reactant. It follows that $[FeL(CH_3CN)(CO)^{2+}] = (6.94-1.25)\times10^{-5} = 5.69\times10^{-5}$ M. An analogous calculation can be performed at each time in the table. This enables a plot of $[FeL(CH_3CN)(CO)^{2+}]$ versus time to be constructed.

Find the concentration of $FeL(CH_3CN)(CO)^{2+}$ from the curve at t = 45 minutes. Divide this by the initial concentration of reactant to find the fraction remaining at 45 minutes. Then multiply by 100. When t = 45 min, the curve indicates that $[FeL(CH_3CN)(CO)^{2+}]$ has fallen to 5.7×10^{-6} M. The fraction of reactant remaining is thus

$$5.7\times10^{-6}/6.94\times10^{-5} = 0.082\text{, or 8.2 per cent}$$

Divide the change in $[FeL(CH_3CN)(CO)^{2+}]$ that occurs over the first 15 minutes by the elapsed time, 15 minutes. This gives the average change in concentration per minute over this time interval:

$$(6.94 - 3.05)\times10^{-5}\text{ M/15 min} = 2.59\times10^{-6}\text{ M/min}$$

Example. The electronic absorption spectrum of a water solution of $CrCl_3$ (chromium(III) chloride) is shown in Figure. The absorption bands, resulting from the Cr^{3+} ion, are in the visible region of the spectrum, and are responsible for the violet colour of the solution. What are the energies of the most strongly absorbed photons? Plot the electronic energy levels of the Cr^{3+} ion on an energy level diagram.

Solution. The most strongly absorbed photons have energies corresponding to the wavelengths at the tops of the absorption bands. The energies, read directly from the absorption maxima in the spectrum, are at 16500 and 23000 cm^{-1}. These are in units of reciprocal wavelength and must be converted to energy units using the relationship

E = hn = hc/lFor the 16500 cm^{-1} photon, E = $(6.626\times10^{-34}$ J/s)$(3.0\times10^{8}$ m/s)(16500 cm^{-1})(100 cm/m) = 3.28×10^{-19} JFor the 23000 cm^{-1} photon, E = 4.57×10^{-19} J, by the same method.

To construct the energy level diagram, we must realise that the observed absorptions correspond to transitions of an electron in the Cr^{3+} ion from the lowest energy state to two higher energy states, one corresponding to each absorption band. The photon energies calculated above give us the gaps between the energy levels.

The energy level diagram should thus show three states. The middle state should be 3.28×10^{-19}J above the lowest one. The highest state should be 4.57×10^{-19} J above the lowest one. The diagram is shown in Figure.

The Correspondence of Colour and Spectrum

Many substances are colored, resulting from absorption of light in the visible region of the spectrum. In addition to making these substances visually interesting, colour is an important physical property, and is used with such quantities as melting point and boiling point in describing the substance. It is important to have a qualitative understanding of the relationship between the colour of light absorbed by a substance and the colour of light that we perceive when we look at the substance.

These colours are *complementary*. We see the light that is NOT absorbed by the colored substance. The colour that we see is that of the light that is not absorbed. Although this statement may contradict what you have previously thought, it nonetheless makes sense.

Photons that are absorbed by a colored substance cannot reach the eye to be observed. The photons reaching the eye are those that are NOT absorbed. White light is made up of the three primary colours, red (R), green (G), and blue (B). The relationships between and combinations of these three colours are best shown in the form of the colour wheel.

The wheel is set up with the primary colours at 12, 4, and 8 of a clock face. The complements of the primaries are shown directly opposite the corresponding primary. Thus cyan, the complement of red, is at 6 oclock. Cyan, C, is a combination of equal amounts of B and G, which occur to either side of it. Similarly, magenta (R + B) occurs opposite its complement, green; and yellow (R + G) appears opposite its compement, blue. A substance that absorbs light of a particular colour appears to have the colour opposite the absorbed colour on the wheel. Thus, if red is absorbed, cyan is seen. Figure shows the correspondence between colour and wavelength of light, which may be seen by viewing white light through a simple device called a *spectroscope*.

The colour of a substance may be understood in terms of its electronic absorption spectrum. Thus if a substance has an absorption band at 550 nm, in the green region of the spectrum, it absorbs green photons, transmits red and blue, and should appear to us to be magenta (R + B).

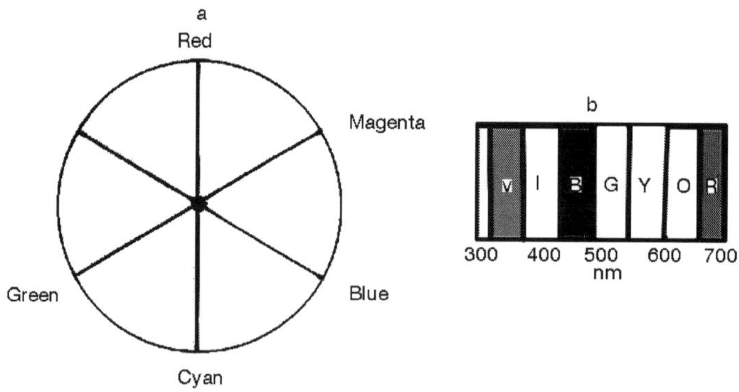

Figure: *Colour*

Relationship of Structure to the Electronic Absorption Spectrum

It is important to have some idea of the types of molecular electronic structural features that are likely to give rise to absrption bands in the near-IR (1100-750 nm), visible (750-400 nm), near-UV (400-190 nm), and far UV (shorter wavelength than 190 nm) regions of the electromagnetic spectrum. We will make some broad statements for which you can expect many exceptions.

The purpose of the statements is to enable you to make reasonable predictions about electronic transitions in molecules. In making predictions, of course, we run the ris of being wrong. But an incorrect prediction based on sound reasoning is far preferabl to no prediction (and no reasoning) at all. You are thus encouraged to be fearless in making predictions.

The near-IR and Visible regions (1100-400 nm). Several structural features are notable for generating light absorption in these regions. First, compounds containing *transition metal ions* (ions of the metals found in the d block of the periodic table) show an interesting rainbow of colours.

These ions are known for their ability to form *complexes*, compounds in which up to six molecules or ions are covalently bonded to the transition metal ion. In this context, the molecules or ions are called *ligands*. In theory, any molecule or ion with at least one lone (nonbonding) electron pair can serve as a ligand by using its electron pair to bond to a transition metal ion. In practice, molecules containing oxygen (2 lone pairs), nitrogen (1 lone pair), sulfur (2 lone pairs), phosphorus (1 lone pair), or halogen (3 lone pairs) frequently serve as ligands.

Table gives formulas and colours for a collection of complexes containing the Co^{3+} ion.

Table: Complexes of Co^{3+}

Complex	Color
$[Co(NH_3)_6]Cl_3$	yellow
$[Co(NH_3)_5Cl]Cl_2$	purple
$[Co(NH_3)_4Cl_2]Cl$	green
$[Co(NH_3)_4Cl_2]Cl$	violet

Without going into too much detail, the colours of transition metal ions are a consequence of the partial filling of the d valence orbitals with electrons. When ligands bind to the transition metal, the d orbitals are caused to have different energies, so that the d electrons can undergo light-promoted transitions between them. The energy spacing between the d orbitals is influenced by the chemical identity of the ligands; thus different ligands coax different colours from the same metal ion. These so-called *d-d transitions* are typically quite weak, with molar absorptivities in the range 0.1-100 $M^{-1}cm^{-1}$ Second, organic molecules with extended *conjugation* (alternating single and double bonds) often absorb visible light. *Acid-base indicators* have structures of this type. Extended conjugation normally leads to quite intense colour (strong absorbance, large molar absorptivity), in contrast to transition metal complexes, whose colours are relatively pale. The *porphyrin ring*, which occurs in many biologically-important molecules, has benzene-like resonance involving its double bonds. It bsorbsintensely in the visible region. Electronic absorption spectral data for a number of compounds of this type ae presented in Table.

Table: *Spectral Data for Conjugated Species*

Molecule	*Absorption Wavelengths and Molar Absorptivities,* $\lambda_{max}(\varepsilon)$
2-nitrophenol, $C_6H_4(OH)(NO_2)$	279 (6600) 351 (3200)
2-nitroaniline, $C_6H_4(NH_2)(NO_2$	283 (5400) 412 (4500)
Acetophenone, $(C_6H_5)(CH_3C{=}O$	240 (13000) 278 (1110 319 (50)

Third, species (molecules r ions) in which an atom in a high oxidation state is bonded to species with lone pairs o electrons (such as oxide ions and halide ions) are often colored, due to the transition of an electron from one of the lone pairs to the electron-deficient high-oxidation-state atom. Such transitions are called *charge transfers*, because they essentially transfer negative charge (an electron) from a region of negative to a region of positive charge.

The permanganate ion, MnO_4^-, is a good example of a species whose colour is due to a charge transfer transition. The anion is intensely purple both in the solid state and in aqueous solution. Table givesspectral data for charge transfer absorptions in a number of species.

Table: *Spectral Data for Charge Transfer Absorptions*

Species	*Absorption Wavelengths*
MnO_4^-	350 (1130) 565 (1270)
$Fe(CN)_5(C_5H_7N_2)^{2-}$	660 (10^4)

Near UV region (400-190 nm). Many organic molecules with double bonds, particularly those in which one of the double bonded atoms has at least one lone pair, absorb radiation in the near UV region.

Examples include acetone, with a carbon-oxygen double bond and 2 lone pairs on oxygen; benzene, with 3 conjugated double bonds; acetonitrile, with a carbon-nitrogen triple bond and a lone pair on nitrogen; and SO_4^{2-} (an inorganic ion), with sulfur-oxygen double bonds and lone pairs on oxygen. Table presents absorption data for a number of common molecules of this type.

Table: *Spectral Data for Species Absorbing in the Near-UV Region*

Molecule	*Absorption Wavelengths and Molar Absorptivities, $\lambda_{max}(\varepsilon)$*
Acetone $(CH_3)_2C{=}O$	279 (13)
Nitromethane CH_3NO_2	275 (15)
Phenol, C_6H_5OH	210.5 (6200) 270 (1450)
Aniline	230 (8600) 280 (1430)

Far UV region (< 190 nm). Molecules containing strong sigma bonds and no lone pairs tend to absorb in this region. The only promotable electrons are those forming the single bonds, and they must receive a large amount of energy in order to undergo a transition to the lowest energy empty orbital of the molecule.

Thus molecules such as ethane, C_2H_6 and other *hydrocarbons* have no absorptions at wavelengths longer than 190 nm. Strongly single bonded molecules with lone pairs on eetronegative atoms also tend to absorb only at very short wavelength (high energy).

Thus water, HF, NH_3, and the lke have no absortions in the visible or near-UV regions of the spectrum. Table shows some typical far-UV data.

Table: *Spectral Data for Species Absorbing in the Far UV Region*

Molecule	*Absorption Wavelengths and Molar Absorptivities, $\lambda_{max}(\varepsilon)$*
Methanol, CH_3OH	177 (500)
Diethyl ether, $(C_2H_5)_2O$	188 (1995) 171 (3981)
Water	167 (7000)
Ethane, C_2H_6	135

The foregoing discussion should give you an idea of the importance and utility of spectroscopy. The standpoint of theory, have been elucidated largely via spectroscopic methods.

Supplement 1: ^{13}C Nuclear Magnetic Resonance Spectroscopy

In the early days of NMR (the 1950s and 60s), the proton was the only nucleus that could be readily observed in the NMR experiment. It was every carbon chemist's dream to directly observe carbon atoms, but the major isotope of carbon, ^{12}C, has no nuclear spin.

The ^{13}C nucleus has spin 1/2, like the proton, so was in principle observable by NMR. The difficulty is that only one carbon atom in 100 is carbon-13, meaning that the number of nuclei absorbing radio-frequency photons would be very small. Early on, the absorptions were indeed too weak to be observed. In the intervening years, however, tremendous improvements in NMR hardware and methods have been made, so that now the chemist can routinely obtain the NMR spectrum due to ^{13}C atoms in a sample.

The characteristics of ^{13}C NMR spectra are fundamentally similar to those of proton spectra. That is, a signal is obtained for each structurally different type of carbon atom; the positions of the signals are measured in ppm with respect to the ^{13}C signal for TMS; the signals tend to be further downfield for C atoms attached to electronegative atoms like Cl and O; and it is possible to observe spin-splitting of the ^{13}C resonances by protons that are directly attached to the observed carbon. However, there are some differences in detail between ^{1}H and ^{13}C spectra. First, the chemical shift range exhibited by carbon is more than 10 times larger than that of protons. Thus although most proton resonances occur within 10 ppm downfield from the TMS resonance, carbon resonances can range more than 100 ppm from the TMS ^{13}C signal. Second, although there is still one signal from each different type of carbon atom, the intensities of the signals do not provide information about the number of carbon atoms generating the signal. Recall that this was one of the very useful aspects of proton NMR; it is not useful for carbon. The reasons for this are complex and we need not bother with them. Third, we expect the signal from a particular type of carbon atom to be split into subsignals by protons attached to the carbons.

Indeed such splitting can be observed, and it follows the usual rules, but most often chemists choose NOT to observe it. Instead, they set the NMR instrument to electronically *decouple* the proton spins, so that only a single line is seen for each chemically distinct type of carbon atom. Fourth, ^{13}C spectra are often obtained by dissolving the compound of interest in *deuterated chloroform*, $CDCl_3$. The carbon atom of the solvent gives rise to an NMR signal that is split into three subsignals due to coupling with the deuterium (heavy hydrogen) atom. This triplet of signals always falls at about 77 ppm downfield from the TMS signal. It should be ignored when interpreting ^{13}C NMR spectra.

The ^{13}C NMR spectrum for chloropentane. The spectrum has the following features:

- It shows a TMS signal at 0 ppm, a solvent signal at 77 ppm, and 4 additional signals, all singlets.

- The signals all fall within 80 ppm of the TMS signal.
- The 4 singlets have different intensities (heights).

The spectrum tells us immediately that there are 4 distinct types of carbon atom in the molecule, although it does not tell us how many of each. Figure shows several possible structures (isomers) for chloropentane. Of those shown the observed spectrum could only arise from options D, E, and G, all of which have 4 carbon atoms in structurally different environments. Without further information, we cannot say which isomer we have. However, suppose that we then re-run the spectrum, this time allowing the attached protons to couple and split the ^{13}C signals, to obtain the result in Figure. The small letters in the spectrum indicate the extent of splitting of each signal: s for singlet, d for doublet, and so on. Of structures D, E, and G, only D is consistent with the observed proton couplings. It should produce quartets from carbons 1 and 4, each coupling with 3 protons; a triplet from carbon 2 coupling with 2 protons, and a singlet from carbon 3 coupling with no protons. Structure E is expected to produce quartets from carbons 1 and 4 and doublets from carbons 2 and 3, while G should show a quartet from carbons 1, 2 triplets from carbons 3 and 4, and a doublet from carbon 2. Let's look at a couple more examples.

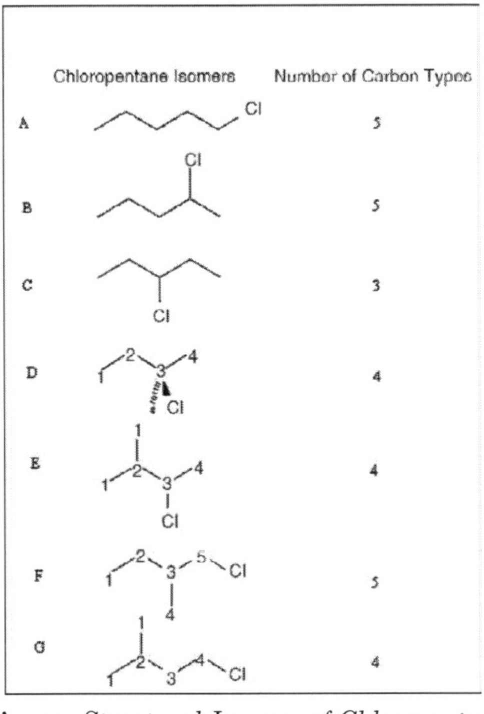

Figure: *Structural Isomers of Chloropentane*

Example. The structure and proton NMR of isobutyraldehyde are shown in figures, respectively. Predict the appearance of the ^{13}C NMR spectrum of this molecule.

Solution. The structure shows 3 types of carbon atom, so 3 signals are expected. If coupling to protons is allowed, the signal from the CH_3 carbons should be a quartet, that from the carbonyl carbon should be a singlet, and the final carbon should produce a doublet. The signal from the carbonyl carbon should be quite far downfield due to the proximity of oxygen.

Example. A compound of formula C_6H_{12} produces the ^{13}C spectrum. What is the molecular structure of the compound?

Solution. Possible atom arrangements corresponding to the molecular formula are shown in figure. Clearly there are quite a few of them! The structures are expected to produce the following numbers of ^{13}C signals:

A – 6
B – 6
C – 3
D – 6
E – 6
F – 5
G – 5
H – 5
I – 6
J – 5
K – 2
L – 1
M – 4

Because 4 signals are observed in the spectrum the structure must be M, which is called methylcyclopentane. If coupling with attached hydrogen atoms were allowed, we would expect to see a quartet, a doublet, and 2 triplets. You might want to predict the splitting pattern that would be observed for each of the other possible structures. It may have occurred to you to wonder whether coupling of one ^{13}C nucleus with another can occur, and whether it is observable in the NMR spectra. In fact, it is not observable; perhaps we should be thankful for this, as it would greatly complicate things! We should spend a bit of time to understand why this type of coupling is not seen.

As said above, the natural abundance of ^{13}C is about 1 per cent (actually 1.11 per cent, but we use the simpler number because calculations are easier). This means that 1 carbon atom in a hundred has mass number 13. We can use this fact to determine the probabilities that varying numbers of ^{13}C atoms will be found in a molecule with, say, 4 carbon atoms. Let's begin with the probability that a particular molecule will have NO ^{13}C atoms. Here is the question we are asking: "What is the probability that carbon 1 is ^{12}C and that carbon 2 is ^{12}C and that carbon 3 is ^{12}C and that carbon 4 is ^{12}C?" In probability mathematics, and means multiply the probabilities for the events that it links. Thus the probability that we seek is $(99/100)\times(99/100)\times(99/100)\times(99/100) = (0.99)^4 = 0.96$. So of 100 of these molecules, 96 will have no carbon-13 atoms in them! The remaining 4 will have between 1 and 4 carbon 13 atoms. Next, let's calculate the probability that a molecule contains four carbon-13 atoms by calculating $(1/100)^4 = 1\times10^{-8}$. This is one atom in 100 million, such a small number that we can ignore this possibility. Much more likely is that a molecule will contain a single carbon-13 atom; this situation can be shown to have probability of about 0.039. The probability that a molecule will contain 2 carbon-13 atoms next to each other, a situation required for spin-spin coupling, is 2.9´10^{-4}. Only 3 molecules in 10000 qualify! Thus the intensity of the NMR signal from the molecules containing one carbon-13 atom is expected to be more than 100 times greater than the intensity from molecules with two neighbouring carbon-13 atoms; the latter is so weak that it will not be seen. Only coupling with attached protons is significant in carbon NMR.

The modern chemist, armed with the techniques of NMR, IR, and UV-visible spectroscopy, is able to discern the detailed structure of just about any molecule of interest. Even large biological molecules yield to some of these techniques. We can be confident that in years to come,science will develop yet other ways to exploit the interaction of light and matter to reveal nature's details at the molecular level.

Supplement 2: An Intuitive Development of Beer's Law

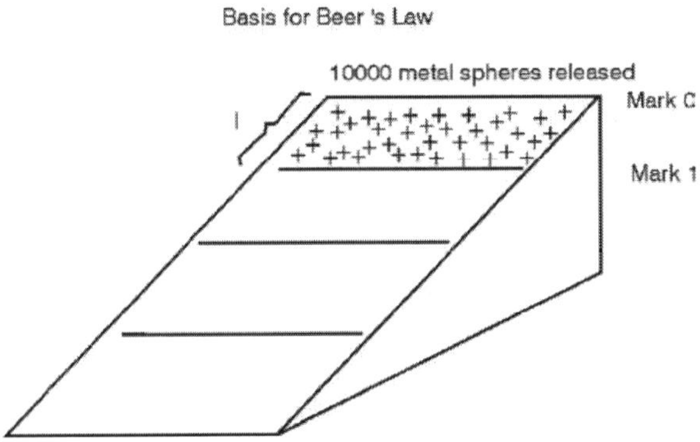

Figure: *Beer's Law Inclined Plane*

It is possible to arrive at Beer's Law using a simple macroscopic physical system consisting of an inclined plane perforated with holes at a certain density, and divided into four sections of equal length. The plane is pictured in figure.

Suppose that 10000 identical small steel spheres are released at the top of the plane, and that we observe that 9000 arrive at the end of the top section, having avoided falling into one of the holes.

Based on this information, how many spheres will arrive at the bottom of the plane? Two approaches seem reasonable on the surface. The first is to assume that the same number of spheres will fall into holes in each section. If 1000 fell in the first section, then a total of 4000 will fall along the entire length of the plane, leaving 6000 to reach the bottom.

The second approach is to assume that the number of spheres that fall into holes in a given section is proportional to the number that enter the section; i.e., that the same fraction of spheres will disappear in each section. By this reasoning, we predict that 8100 spheres will survive the first 2 sections, 7290 will survive through sections, and 6561 will reach the bottom. This second line of reasoning proves correct: the number of balls that survive a section is proportional to the number of balls that start into the section. We write this in equation form: $N_n = k \times N_{n-1}$

This equation states that the number of spheres arriving at mark n is proportional to the number of spheres that leave the preceding mark, n-1. k is a proportionality constant that must have a value between 0 and 1; it has the value 0.9 for our example above. We can also express the equation in terms of the change in the number of balls:

$$DN = N_n - N_{n-1} = k \times N_{n-1} - N_{n-1} = N_{n-1} \times (k-1) = -K \times N_{n-1}$$

Note that because k < 1, the quantity k-1 is negative, consistent with the fact that balls are being lost and therefore that the change in the number of balls defined in equation is also negative. We have therefore replaced k-1 with –K, where K is a new, positive, constant.

Here are two questions that we now consider: How does the value of K depend on the length of the interval, *l*? How does the value of K depend on the density of holes, r? Intuitively, we expect that a longer interval will result in more balls being lost. Thus K increases as *l* increases. Also, we expect more balls to be lost if the density of holes is increased. So K increases as r increases. In equation form, this becomes:

$$K = \zeta \times l \times r$$

where z is a constant representing the number of balls lost per unit density of spheres per unit length of section. Substituting 4S-3 into 4S-2 gives

$$: \Delta N = -\zeta \times l \times \rho \times N$$

Assuming an interval of infinitesimal length will result in an infinitesimal loss of spheres:

$$: dN = -z \times r \times N \times dl$$

Separating the variable, N, from the variable, l, gives

$$: dN/N = -\zeta \times \rho \times dl$$

Finally, integration provides

$$: \ln(N/N_0) = -\zeta \times \rho \times l$$

Translating to base-10 logarithms gives:

$$\log(N/N_0) = -(1/2.303) \times \zeta \times \rho \times l = -e \times \rho \times l$$

Of course this equation can be expressed in exponential form as $N/N_0 = 10^{-e \times r \times l}$.

In words, this equation states that the fraction of balls surviving an interval decreases exponentially with the length of the interval and the density of holes in the interval.

Let's now make a direct analogy between our inclined plane and a solution of a substance that absorbs light at some wavelength.

We think of l as the length of solution through which the light beam must pass; r as the density (concentration) of absorbing molecules in the solution; the small spheres as photons of light; and N/N_0 as the fraction of light transmitted by the solution:

$$\log(I/I_0) = -e \times M \times l$$

We have replaced the density of holes with the concentration of the solution in moles absorbing substance per liter, and the number of spheres with the intensity, I, of light.

The intensity is simply the number of photons per second arriving at a particular location, so is a close analog of N. We now define a quantity called the absorbance, A, as $A = e \times M \times l$ so that equation becomes:

$$\log(I_0/I) = A.$$

A represents the amount of light absorbed (the number of spheres lost). Note that when all light is transmitted by a sample, $I_0/I = 1$ and $A = 0$. When 1/10 of the light is transmitted, $I_0/I = 10$ and $A = 1$. Thus A goes up as the fraction transmitted goes down. The equation $A = e \times M \times l$ is, of course, Beer's Law.

Atomic Structure and Bonding

Macrophysics and Microphysics

Modern physics accepts the atomic structure of matter. Matter, palpable and visible to us, is made up of large numbers of minute bodies, called atoms, which can neither be seen nor felt. Greek philosophers had suspected the existence of atoms, and this concept taken over from the Greeks has played an important part in scientific investigation and development since the beginning of western scientific research. But it was not until this century that what had been supposition, speculation or fantasy was raised to the rank of solidly established scientific knowledge confirmed by direct experimental proof. It can even be said that ours is the century of atomic research in physics; and a survey of the ideas of modern physics should first of all deal with the investigation of atoms.

The moment, let us consider a piece of homogeneous matter capable of being subdivided. A piece of copper or a piece of rock crystal can be broken and split up into smaller and smaller pieces, and the fragments continue to show the characteristic properties of the copper or the quartz. Even if certain limits existed for the practical accomplishment of such a subdivision—it is difficult to break up into smaller parts quartz which has already been ground up to dust—yet this rough experiment presents no criterion for the existence of a definite, clear limit to further subdivision. The dust particles still differ in size and shape among themselves.

Thus one might conclude, that—even though our tools are ineffective—nature presents no obstacle to halving the smallest dust particle once more, and to halving this half again, and so on... without a limit. Today we know with certainty that that is false. But very refined methods are required to attain the limit to the continued division of matter—in other words, for the experimental determination of the atomic structure of matter. But, as previously stated, such methods have been available for only a few years.

Thus there are large and rich fields of physical research in which the atomic structure of matter is not at all recognizable. Naturally no information about atoms can be expected from such research, but then no knowledge of atoms is necessary, as long as investigation is limited to these fields. It is customary to speak of macroscopic physics (or macrophysics) when referring to those investigations in which the presence of atoms is not discernible; on the other hand research which penetrates into the atomic detail of matter is denoted as microphysics.

This review is devoted in large part to a discussion of microphysics—but since only a thorough comprehension of macrophysics can make possible an understanding of microphysics, we shall begin with a discussion of the former. Moreover, it was in our century that macrophysics reached its final ma turity. It was in these recent developments that certain philosophical theoretical thought processes were developed and clarified, making it possible to reconcile the surprising and paradoxical results of microphysical experiments, which previously were difficult to understand.

We are beginning with very simple things. Despite their simplicity, however, they merit careful and thoughtful consideration.

The Laws of Falling Bodies

The Greeks had already established the laws of statics, of mechanical equilibrium of levers and of fluids. The essentials of these laws had been clarified. On the other hand, dynamics, the theory of the mechanical laws of motion, was first established by Galileo. It is not within the province of this review to trace the historical development of Galileo's investigations; although for a deeper spiritual historical understanding of modern science, closer consideration of the historical origin of modern physical thought—up to the end of the Middles Ages—would be very valuable. We shall simply consider Galileo's laws of falling bodies in their finished form. The simplicity of these laws and the ease with which they can be interpreted today should not deceive us as to how tremendous was his mental capacity. To judge his contributions fairly, the historical-cultural background and the ideas—so completely changed and readjusted since then—prevalent in pre-Galilean times should be considered.

Let us consider the motion of a falling or pro jected body. Simple, everyday experience indicates that a feather falls more slowly than a piece of lead; and Aristotle had taught generally that light bodies fall more slowly than heavy ones. Galileo's main contribution is that he recognized the possibility of extracting an obscure but simple law from the amazingly complicated motions of real projected bodies; he considered falling motion in a vacuum. It is possible to prove that in an evacuated tube a feather falls as quickly as a piece of lead—this experiment is part of the course of study in schools today. But Galileo's contemporaries still considered the idea of a vacuum as impossible, as nonsense.

In those times it required far greater strength of intellectual abstraction than we might expect today, to recognize that it was possible to separate (mentally) all the associated events from the motion of falling bodies. We know today that the effects which originate from atmospheric pressure and the influence of the wind can lead the motion of a falling feather to proceed entirely differently from that of a falling stone.But it is possible to extract from all these complex and varied motions an ideal form which is to be regarded as the motion of a body in a vacuum.

This "ideal" projectile motion is wonderfully simple compared to the actual motions of falling bodies or projectiles:

- Ideal projectile motion is always a plane motion (proceeding within a fixed plane perpendicular to the earth's surface).
- It proceeds independently of the mass, size and shape of the projected body; it simply depends upon the magnitude and direction of the initial velocity.

It is not influenced by any rotation of the projectile around its own centre of gravity; even if the body rotates around it, this centre of gravity moves according to the law of ideal projectile motion.These properties do not properly describe the motion of a real bullet, which is not exactly a plane motion, and which depends very substantially on the mass, size and shape of the projectile as well as on its spin (rotation).

Here we witness one of the historically most important examples of physical natural scientific thought. The phenomena presented to us directly by nature are so varied and complex that it is impossible for our narrow minds to grasp them individually. We must mentally divide these natural occurrences into their simpler components; we must consider results of artificially produced extremely unnatural conditions instead of those offered through direct experience. Thus we uncover the ideal cases, which can be subjected to more precise, considered treatment. These results in turn establish criteria for the evaluation of the real events, in each case different from the ideal to a greater or lesser extent.

Now let us consider the laws of ideal projectiles more exactly. For this we want to view the simplest case, namely vertical rectilinear fall. The body, first retained at rest and then released, falls downwards—as long as it continues to fall with ever increasing velocity. At the instant of release, the velocity is exactly zero, but thereafter it increases steadily. This increase does not proceed in intermittent, rapidly repeating, sudden jumps, but rather in an uninterrupted, continued growth, in which there are neither pauses nor sudden discontinuities. Mathematicians express it very clearly when they say that the velocity increases uniformly.

In such a case, when the velocity is changing uniformly and constantly so that it is never the same at two different points in time, no matter how closely they may follow one another, what does velocity mean? When a body moves with uniform velocity it is clear what velocity means—the distance traversed is divided by the time of travel. We commonly say that a body traveled at a velocity of one meter per second or that an auto traveled at seventy kilometers per hour. But how are we to understand and define velocity abstractly if it doesn't remain fixed throughout even a fraction of a second? In Galileo's times this question presented very great difficulty; the abstract and mathematical tools for solving it were still lacking. In an evaluation of Galileo's contributions, this fact, too, must be remembered.

Galileo had been able to dispose of these questions completely for ideal projectile motion; it only became evident later how great an achievement this was. Newton cleared up the concept of velocity (and acceleration) quite generally, for any motions. For this purpose he had to establish an entirely new branch of mathematics, the so-called differential calculus. Differential calculus is undoubtedly the greatest creation achieved by western mathematicians. In this Leibnitz stands beside Newton as the founder—although his considerations did not emanate from mechanics, but rather from geometry.

Naturally both of them, Leibnitz and Newton, referred back to forerunners and trailblazers; in the problems of their time and in their natural scientific and mathematical investigation there lay an inevitable compulsion towards this direction of thought. But what they produced by bold statements occupied generations of mathematicians after them and stimulated further great results.

This new mathematics, which is an original creation of western thought, not anticipated by the Greeks, deals with precisely those questions which confront us when we inquire about the exact meaning of the "concept" of velocity in the case of constantly changing velocity. The function of differential calculus is to express clearly mathematically quantities that describe constant, fluid, continuous change.

The exact definition of velocity can be expressed as follows. First, it is clear what is meant by the average velocity within a given time interval—the distance traveled in this time is divided by the length of time. Now, in order to obtain the exact velocity for a given point in time we choose a small time interval (including this point in time) and determine the average velocity associated with this interval; it will, if the time interval chosen is small enough, give an approximately correct value for the exact velocity desired. This value can be improved by replacing the time interval with one half as large; and then this improvement can be repeated. It is frequently possible to repeat it mentally at pleasure, and thus with unlimited approximations to approach ever more exactly the precise value desired.

This, therefore, is the definition of the concept of "velocity". The problem is no simpler if one wants to obtain the results quantitatively, and for this purpose wants to sharpen the concept so that it will make a mathematical evaluation possible -when one does not want to be content with a solely emotional, indefinite application of the concept. But one need not fear that this has led to a hopeless, practically insoluble problem in calculation. On the contrary, there are ingenious considerations (and their cultivation is the inherent content of differential calculus) which make it possible to establish in a specific form of motion the limiting value which we must take as the basis for the definition of velocity.

Galileo had already answered these questions for ideal falling motion. He had recognized that for this ideal motion, dissociated from air resistance and all secondary influences, a wonderfully simple law exists; namely, velocity increases proportionally with time. Thus, after double, triple, etc., time of fall it is exactly doubled, tripled, etc.

This law extends also to general ideal projectile motion. If we throw a body vertically upwards or downwards, its downward velocity will still always increase in proportion to time—which in the case of upward motion naturally means a corresponding decrease in the upward velocity. If it is thrown obliquely, the body's height measured vertically to the earth's surface (in ideal projectile motion) changes with the passage of time exactly as if it also were a vertical fall.

Simultaneously, the horizontal distance from the starting point increases proportionally with time; or expressed differently, in the horizontal direction the motion proceeds with fixed velocity. Mathematical consideration of these determinations yields the fact that the projectile path (trajectory) is a parabola.

Force and Motion

The laws of generalized ideal projectile motion formed another example of how scientific thought made natural processes comprehensible through resolution into simpler components.

We conceive of parabolic motion as simultaneous execution of two different motions, vertical and horizontal. It is also possible to study the horizontal motion isolated from the vertical -for this all we need is a horizontal rail, or perhaps a plane ice surface, on which the body can slide. Then for the ideal limiting case (*i.e.*, in the case of not only the absence of air resistance, but also of any friction accompanying the sliding) we reach the conclusion that the body moves in the horizontal direction with constant velocity.

Galileo accepted the certainty of the spherical shape of the earth, at that time not a very old concept; and he could represent the earth as moving freely through space on the basis of the Copernican theory which he accepted. Thus he knew that the motions horizontal and vertical or above and below are only relative.

And from this the possibility of generally recognizing the basic form of motion, free from external influences, in the ideal, resistancefree motion of a horizontally sliding body became evident. Newton, the first one to state this clearly, expressed his famous law of inertia as follows: a body moving in a vacuum unobstructed and free from external forces moves in a straight line with constant velocity.

Falling motion must thus be conceived as a deviation from the invariable behaviour of a body, for which the cause is to be sought in an attractive force emanating from the earth. Newton clarified for all cases the manner in which a force acting on its centre of gravity alters the motion of a body -the force manifests iteslf in a change of velocity, an acceleration. In vertical free fall the acceleration is constant—whereby we define acceleration as the increase in velocity divided by the duration of the fall. For complex motion, in which the acceleration is no longer constant but is subjected to continuous changes the same considerations are necessary as were introduced for the concept of velocity. Newton's new mathematics, differential calculus, mastered all these problems with one stroke.

If we limit ourselves to rectilinear, uni-dimensional motion, then according to Newton the acceleration is exactly equal to the force divided by the mass of the accelerated body. It depends on neither size nor shape, nor material, colour nor temperature of the body, but on its mass alone.

The laws of three-dimensional motion are analogously simple and general. Here it is simply neces sary to resolve the motion into three components, acting in mutually perpendicular directions, whose simultaneous execution yields the total effective motion. If the force acting on the body's centre of gravity is resolved correspondingly, the above Newtonian law is valid for each of the three uni-dimensional portions of the whole process. This knowledge made possible a clear definition of the concept of physical causality. The general notion that nothing happens unless a definite cause exists for it was elevated to a quantitative law: a definite force acting on a body imparts a definite acceleration to that body.

This point merits further consideration. To really understand the significance and importance of the ideas which constitute physics, we must always be ready to refer back to previous concepts. Only in this way can we realise how revolutionary such thoughts were at one time, although they have become quite familiar and conversant to us.

As the most prominent feature of natural-physical research we recognize the search for quantitative, mathematically comprehensible laws. We must remember that extreme idealization of natural events is necessary to render this search successful. Since, by eliminating air resistance, Galileo attained the process of ideal fall, he was able to uncover mathematically simple, beautiful, exact laws; and conversely, these laws show that he looked for the idealization in the proper direction. Naturally calculations on the projectile trajectories of modern weapons do not yield these ideal projection laws at all. Such calculations require consideration of air resistance—which also is introduced in a more or less idealized form. Ballistics, which deals with these problems, is a special, highly developed science whose outstanding practical significance needs no special emphasis. But the science of ballistics had nothing new to add for the general development of physical science, for the development of physical thinking and ideas.

The idea of continuity, which attained its mathematical form in differential calculus, is important for the clear understanding of motive processes. We also want to make it clear immediately that this continuity of natural events—"natura non facit saltus"—was already evident in the elementary fact that it was at all possible to speak of a definite trajectory of a moving body.

A body cannot reach one place from another by jerks, suddenly disappearing here and emerging again there; it must describe a continuous connected path between the two. But why is that necessary? We know from experience that it always is that way, but is there a logical necessity that it cannot be otherwise? These questions are not idly posed: we shall never be able to understand microphysics unless we have carefully examined such questions.

If we continue to consider macrophysics, it must be realised that in this sphere the principle of continuity is valid without exception. When a shell explodes, the proposition holds that individual fragments can only change their positions through continuous motion, rapid as it may be. Or in an automobile accident—the law obtains that each body can only change its velocity continuously, never with "discontinuous", complete suddenness.

Also, the quantitative, mathematical definition of mechanical causality, in which Galileo's and Newton's knowledge culminated, is inseparably bound up with the concept of continuity. For, it is through Newton's law—"force equals mass times acceleration"—that constant changes in velocity are traced back to the forces which cause them.

Relative Motion

The knowledge expressed in the law of inertia and in Newton's definition of the operation of force was diametrically opposed to earlier views. Previously the problem had been viewed just in reverse—explanations were being sought for the fact that a hurled stone retains its velocity after being released from the hand. It was considered natural and understandable that it must lose its velocity if the impulse is lacking to maintain it. For everyday experience does indicate that a wagon, for example, which is supposed to move uniformly along a straight path, does require a continuous uniform force from the beast drawing it.

The now recognized fact that, exactly opposite, an uninfluenced body retains its velocity without any change—the wagon on the road is not a valid example of this because it is retarded by the action of friction—and that quite generally only acceleration, and not velocity, is determined directly by the acting force, is related to certain very significant problems, which were not fully solved until this century. For the moment these problems are pertinent only so far as they refer to the mechanics of Galileo and Newton.

Opponents of the Copernican theory of the motion of the earth around the sun had introduced the following objection: if a gun is shot vertically upward, the shot falls down close to the gun. But, they said, according to Copernicus, since the earth has in the meantime moved along some distance, the shot should have fallen down in an entirely different place. This was impressively demonstrated in an experiment by Gassendi: he dropped a stone from the tip of the mast of a rapidly traveling boat. It fell down below near the mast, not on the stern of the boat nor behind the boat in the water.

Galilean-Newtonian mechanics explained this result without more ado. And, since generally it predicted suitable results for all analogous experiments, it lent an indispensable contribution to the justification of the Copernican theory. To our present thoughts, which accept the ideas of Galileo, Newton and their associated mechanics as a basis, the process is clear immediately: the stone, first held in place at the tip of the mast and then released, received the same velocity in the horizontal direction as the boat; and it retained this horizontal velocity (as long as the effect of air resistance was negligible) so that as it fell downwards it moved along with the ship in the horizontal direction.

From the Newtonian laws of mechanics it is obvious that if a boat (or train) is traveling in a straight line with constant velocity, one cannot determine on the inside of a closed cabin of the ship whether or how fast the boat (train) is traveling. Although naturally it can be deduced from the altered rocking of the boat or swaying of the train, here we are considering the ideal case of a vehicle moving along smoothly without rocking or bumping.

An object dropped in the cabin will fall vertically to the floor, precisely as it obeys the law of falling bodies on land. Now let us consider that an observer standing on shore can see inside of the cabin. The author desires to establish that the falling object is still subject to the laws of falling bodies, although this observer would not see it fall vertically, but in a parabola—with a horizontal velocity (as long as the effect of air resistance was that of the boat).

The concept of velocity is relative. The observer on shore views all objects in the cabin with a different velocity than the experimenting traveller; for the witness on land every object moving in the cabin acquires in its motion also the additional velocity of the vessel. Therefore, to avoid misunderstanding we must always append to the word velocity "relative to the boat" or "relative to land". When a boat travels with constant velocity (relative to shore) then the acceleration of any body relative to the ship is always the same as it is relative to land despite the difference between the relative velocities. Consequently the same mechanical laws apply to motion relative to the boat and relative to land. In both reference systems the same accelerations are caused by the forces acting. This discovery, that one can never determine uniform motion of a closed room from its inside with any mechanical apparatus—be it simple or complex—whose method of operation is subject to Newton's laws, is called the principle of relativity. The principle of relativity is, beside the principle of the conservation of energy, the most general and most comprehensive physical law that we know.

As yet we have only established its validity with respect to the laws of Galilean-Newtonian mechanics. But we shall see that its significance extends far beyond this and embraces all fields of physics.

Therefore we can't perceive through any mechanical experiments conducted on the earth that the earth (relative to the sun) glides along thirty kilometers per second, or that the sun (relative to the milky way system) flies along with still greater velocity. And we can't distinguish whether the milky way, which despite its tremendous size is only a small island in space, for its part executes as a unit any uniform rectilinear motion.

The rotation of a body, for example, can be recognized by means of mechanical effects. Anyone sitting on a carousel can tell, even with his eyes closed, that it is turning from the centrifugal force which presses his body against the outer wall—since according to the law of inertia the body would fly out in a tangential direction due to the rotary motion if the outer wall didn't retain it. In the same way we notice a curve within a train; we also notice the acceleration or deceleration of rectilinear motion when starting or stopping. It is only uniform progression which is not noticed, because only there is the acceleration of the body relative to the vehicle the same as that relative to the fixed rails.

In the case of our earth the centrifugal force caused by its rotation is evidenced in the flattening of the poles. It is also possible to measure the earth's rotation directly through a simple mechanical experiment—whereby the Copernican conception of motion in our solar system, developed from astronomical knowledge, was shown to be a necessary result of Newtonian mechanics. This is the famous Foucault pendulum experiment. A weight swinging on a very long thread transferred into an elliptical form of vibration under the influence of the earth's rotation. The measurement of this effect permits the determination, based on Newtonian mechanics, that the earth actually rotates with the same speed that the "fixed star" sky seems to move (oppositely directed); that actually the "fixed star" sky, as Copernicus taught, remains at rest.

Chapter 3

Evaluation of the Atomic Concept

Dalton first recognized and effectively utilized the stimulating force of the atomic concept and its ability to guide future physical research. To him we owe the use of the idea of atoms for understanding basic regularities of chemistry. Chemistry makes a sharp distinction between mixtures of different substances and chemical combinations. A mixture–as, for example, a solution of sugar in water or a mixture of nitrogen and oxygen–can include arbitrary (within certain limits) relative proportions of the constituents. In a nitrogen-oxygen mixture the portion of oxygen can be varied continuously by addition or removal; correspondingly, the mixture exhibits properties somewhere between those of pure oxygen and of pure nitrogen, depending upon the relative concentration.

Also, the mixture can be separated into its constituents through relatively superficial means. In a chemical combination entirely new properties appear, which can be quite dissimilar to those of the constituents. The proportions of the components of a chemical compound are quite stable and cannot be varied continuously. Thus exactly 16 grams of oxygen always combine with 2.016 grams of hydrogen to form water.

Another chemical substance, hydrogen peroxide, can be formed from 2.016 grams of hydrogen and just 32 grams of oxygen. It is striking that exactly twice as much oxygen is utilized in the second case as in the first; and on the basis of the atomic concept Dalton found a clear explanation for such facts, which occur similarly in all chemical combinations. His explanation was as follows: The "molecule" of water, *i.e.*, the smallest possible particle of water, contains only one atom of oxygen; the molecule of hydrogen peroxide however contains two atoms of oxygen, while it contains just as many (two) atoms of hydrogen as the water molecule.

Here we arrive at the distinction between atoms and molecules; we designate the smallest particles of chemical compounds as molecules, which for their part consist of atoms of the chemical elements. According to Dalton's ways of thinking a comprehensive study of proportions by weight in chemical compounds rendered the determination of the relative atomic weights of all the elements possible. The "atomic" weight of oxygen is arbitrarily designated by the value 16, whereupon hydrogen acquires the atomic weight 1.008 and every other element, similarly, receives a definite atomic weight derived from chemical weight relationships. The regularities which the gases exhibit, which in the case of sufficiently small densities assume an "ideal" form, proved very helpful for the clarification of these ideas; for example, in the sense that a gas mass contracts to one-half its original volume when, with constant temperature, the pressure impressed upon it is doubled.

Criteria existed for the hypothesis that not only in compounds but also in a gaseous element such as nitrogen the particles moving at random through space are not necessarily identical with atoms; in nitrogen, for example, each "molecule" of the gas consists of two atoms. In the chemical combination of gases (with constant pressure and temperature) there is a simple relation between proportions by weight and volume relations (proportions by volume) between the chemical compound and the still uncombined constituents.

To fit these facts Avogadro introduced the explanation that equal volumes of all gases measured at constant pressure and temperature always contain the same number of molecules. Based on this "Avogadro's principle" a "molecular weight" can be determined for each gas independent of chemical weight relations.

The molecular weight of a chemically homogeneous gas is equal to the mass of 22.4 liters of this gas measured in grams at one atmosphere of pressure and a temperature of 0° C. Experience shows that values obtained in this way correspond with the chemically defined atomic weights. In many elements, *e.g.*, the metals, this molecular weight is exactly the same as the atomic weight; thus metallic vapors are "monatomic". In the aforementioned nitrogen, however, the molecular weight is exactly double the atomic weight. And in water it equals 18.016 ($2 \times 1.008 + 16$) in accordance with its above-described chemical composition. These ideas are further verified by simple relations of molecular weights such as, "diffusion velocities", or by the characteristic difference between the specific heats of "monatomic" and "polyatomic" gases. None of these, it must be emphasized, are proofs for the correctness of the atomic concept; but they are proofs for its usefulness. The atomic idea describes very plainly a large number of important and comprehensive regularities, thus fadlitating greatly our progress with these phenomena.

Crystallography furnishes evidence of further regularities which are certainly clearly explained by the atomic idea, though they too do not lead to proofs for the reality of the atom. In the crystalline state of matter the atomic concept suggests a regular grouping of the atoms or molecules, alongside of each other and arranged in layers. This view induces important results; on this basis the variety of possible forms of crystals are shown mathematically to be limited very considerably. The completion of this mathematical proposition yields the wonderful result that according to the atomic concept there should be no more nor less than 32 different "crystal classes" which are defined by different properties of symmetry.

Actually this confirms precisely the experience of mineralogists; there really occur in nature all the crystal forms (crystal symmetries) which are compatible with the atom idea, but not one single one which contradicts it. The so-called "law of rational indices" discovered by the mineralogists formed an important adjunct to this original consideration, which is related only to the symmetry properties of different crystal forms. If we imagine the crystal as built up of a regular stratification of layers of atoms, then its outer boundary surfaces could never lie so that they partially cut through the atoms. The limits for the possible positions of the boundary surfaces given by this view are exactly equivalent to the requirements which the law of rational indices places on the position of the crystal surfaces.

Such experiences must necessarily encourage the more and more energetic pursuit of the atomistic notion–which intrinsically was so clear–even though tangible proof for the reality of atoms was still lacking. Actually the theoretical development of the atomic concept was furthered extensively by the most able physicists. We have already mentioned ideal gases and indicated how the gaseous state of aggregation should be represented on the basis of the atomic theory; the molecules of a gas are completely separate from each other; each one moves along with high velocity until it meets another one, and then, in elastic rebound, their direction of motion and speed change.

The continuous impact of countless molecules on the walls of the enclosing vessel produces the pressure of the gas, perceptible to us with our rough tools. These ideas were elaborated thoroughly (Clausius,

Maxwell) and our understanding of the gaseous state of aggregation was developed very clearly and completely. Whereby the proof resulted that Avogadro's principle, first only speculatively suspected, must necessarily be correct if the atomic hypothesis proves at all valid.

The discovery of the energy principle (Mayer, Joule) further strengthened confidence in atomism. It was recognized that mechanical energy, which disappears through friction or the like, reappears in the quantity of heat produced–in such a way that a definite amount of consumed mechanical energy corresponds to a definite number of calories of heat. Conversely, according to the same conversion relation, in steam engines quantities of heat are transformed again into mechanical kinetic energy. This conversion of mechanical energy into heat and its converse can be considered quite consistent, in the sense of "macrophysics". Whereupon one must establish that energy, which has proved indestructible, can assume both the form of a quantity of heat and of mechanical energy. The "macrophysical", "phenomenologic" heat theory thus arrived at is entirely sufficient for answering all the questions encountered in connection with heat engines, and with heat conversion in chemical processes.

But further thought is suggested to give the process of the conversion of mechanical energy into heat a clear interpretation in the light of the atomic theory. In a gas, as was represented above, the concepts of heat and temperature are not applicable to the individual molecules, but only to the gas mass as a whole. In an individual molecule only the kinetic energy, *i.e.*, its velocity of motion, or at best also its rotation, can be changed. Thus, if energy is added to a gas mass, which appears macrophysically as added heat, it can only mean that the average energy of motion of the gas molecules has increased. The exact determination of the connection between the average energy of motion of the gas molecules and the temperature of the gas is a part of the theoretical investigations on the "kinetic theory of gases", which has already been discussed.

Similarly, in a solid body, a crystal perhaps, the temperature and heat energy contained within it are conceived of as mechanical energy of its molecules or atoms. Although in their regular arrangement in layers no one atom can leave its place, fine vibrating motions within the crystal–possibly similar to the fine vibration of a heavily traveled steel bridge–are still possible.

If these vibrations within the crystal are so small that they are not perceptible to us as mechanical motion, we do note the energy of these motions–as heat content which manifests itself in the temperature of the body. (If the atoms contained in each small piece of the crystal move by very small amounts in random directions, the whole crystal appears motionless to our macrophysical senses.) Apropos of such thought processes was Boltzmann's atomistic interpretation of a remarkable regularity in heat phenomena. If we imagine that a planet's motion is interrupted and the planet then is induced to move in the opposite direction with the same velocity, it will retrace its entire elliptical path in the opposite direction after the reversal.

This is an example of the fact that all purely mechanical motions are, as we say, reversible. The upward motion of a stone, thrown from the earth, (in the ideal case) until its-point of reversal is an exact temporal mirror image of the subsequent downward motion. Similar reversibility–which can also be denoted as a physical "symmetry of the positive and negative time orientation"–exists in all purely electromagnetic properties.

But when we move a body over a table surface against frictional forces and heat is generated thereby, there is no reason to expect that in moving the body backwards along the same line the converse is true–that the heat energy will change back to energy of mechanical motion. Or take an electric current which flows through a wire and produces heat–when it flows through the wire in the opposite direction it again produces heat; the heat energy does not change back into electrical. These are examples of irreversible processes. Another example of such a process is the mixing of two liquids; in most cases these intermix by themselves, but this process never runs backward of itself.

A further example of importance to us is the following: if a gas filled vessel is introduced into an evacuated one and is opened there, the gas diffuses throughout the entire available volume. It is not impossible to restore the original conditions; the larger vessel can be evacuated again by pumping and the gas can be compressed into the smaller one.

But this does not constitute an exact reversal–a temporal reflected image–of the first process; and it can be considered that for the restoration of original conditions after any irreversible process a definite "compensation" must be included (Clausius). In the steam engine, which undertakes to convert quantities of heat into mechanical energy, there does not occur a simple reversal of the process of the production of heat through friction, but each steam engine functions according to a scheme similar to the following: a certain amount of heat is removed from a heat container maintained at a high temperature. A fraction of this amount is changed into mechanical work, while the remainder is conducted into a heat reservoir of lower temperature. As "compensation", therefore, for the conversion of heat into work there occurs the transfer of a quantity of heat from a hotter body to a colder one–*i.e.*, therefore a process which nature could perform "irreversibly" itself, since in heat conduction the heat always flows from higher to lower temperature.

The occurrence of such "irreversible" processes among heat phenomena posed a very difficult problem for the "kinetic theory of heat"–tracing back the laws of heat to the mechanics of atoms. For purely mechanical processes were shown to be always reversible, which makes it appear impossible to trace these irreversible heat phenomena back to mechanical properties.

The solution of this difficulty, achieved through Boltzmann's perspicacity, runs as follows: in an aggregation of many similar particles, such as are attributed to every physical system by the atomic theory, there is no point to pursuing the motion of each individual atom or molecule; it is only important to us to observe the average statistical behaviour of large numbers of atoms.

Thus statistical concepts are necessarily introduced into the consideration and we must determine which (defined in a rough statistical sense) events are to be considered as especially probable (thus as occurring very frequently) in comparison with other conceivable processes.

For our example ths signifies that when the small gas filled vessel is opened in the large empty one "it is extremely probable" that the gas will stream out and distribute itself uniformly in the large vessel. Strictly speaking it cannot be definitely known that this will happen; indeed, to view the processes of motion in the gas mass with the same complete certainty achieved for planetary motions one must know quite exactly the position and velocity of each individual molecule at the beginning of the experiment. The fact that, instead of this, we simply denoted the initial condition of the experiment statistically stipulates that we can not predict its further progress with complete certainty, but only with very great probability.

The reversibility of pure physical processes, which according to the atomistic conception is the basis for macrophysical heat phenomena, depends upon the existence of an exact reversal, an exact temporal reflected image for every process–whereby it can happen that the gas mass distributed in the abovementioned large vessel contracts into the smaller one, in an exact reversal of the normal process.

No one can guarantee that this is completely out of the question; but we can show that such an event is extremely improbable–mathematically its probability is expressed by an infinitesimally small number. Thus it becomes clear that despite the fundamental reversibility of atomistic elementary-processes, yet, on a large scale, in macrophysical events, practically irreversible phenomena take place.

Another instructive example of an irreversible process is the mixing of two gases or liquids with each other. According to the kinetic-atomic theory this problem is somewhat similar to the one wherein two types of balls–red and white ones for instance–are thoroughly shaken up in a sack. If the red and white spheres are carefully separated initially, they will become completely mixed up after further agitation. If, then, we continue to shake, it is possible in principle that the original separation of red and white balls will be restored–but practically this possibility is of little moment since it would require shaking for astronomically long periods of time before this "spontaneous separation" could be expected with appreciable probability.

A quantitative measure for the irreversibility of a thermodynamic process (from a purely macrophysical standpoint) can be specified in the form of "entropy"–a quantity which always increases to a maximum in irreversible processes and does not decrease again (Clausius). According to Boltzmann this entropy can now be defined through the atomic concept as a measure of the disorder in an atomic aggregate.

The Limits of the Divisibility of Matter

Everyday experience, which appears to show unlimited divisibility of matter, through simple refinements makes it certain that atoms must be extraordinarily small, and that therefore the number of atoms in a gram of any substance must be enormously large. For an exact measure of this quantity we use the so-called Loschmidt number. As a "gram-atomic weight" of an element we denote a mass of a number of grams equal to the atomic weight; the Loschmidt number is the (equal for all elements) number of atoms in a gram-atomic weight.

Evidence for the divisibility of matter can be obtained by distributing a minute amount of a strong-scented substance (mercaptan) in a large quantity of air and then determining to what degree of dilution the odour can still be perceived; or by dissolving a strongly coloured liquid (eosin) in a relatively large quantity of water and then noticing to what degree of dilution a uniform coloration of the water remains perceptible.

A still more suitable experiment involves the use of a very dilute solution of fluorescein, a substance which fluoresces strongly in incident light. Extremely small volumes of the solution are examined with a microscope to determine to what degree of dilution a spatially uniform fluorescence can be detected. From such observations it can be inferred (Perrin) that the Loschmidt number is definitely greater than 10^{21} (a 1 with 21 zeros after it).

But quite simple, daily experiences occur that indicate the limits which its atomistic structure places on the subdivision of matter. These involve utilization of very thin membranes of matter. Gold leaf, which is used for gilding purposes, is hammered down to an exceptional thinness–when held against the light it exhibits a greenish translucence–of about 100 millimicrons (a million millimicrons make a millimeter); and yet this thickness is still far removed from the ultimate limit. But there exist direct indications for the atomistic constitution of matter in commonplace, familiar soap bubbles, often formed when washing by a film of soapy water stretched between the thumb and index finger.

These bubbles iridesce in variegated colours, whose diversity and rapid shifting is evidence of the variations in the thickness of the bubble with position and time. Small dark spots appear in a thus coloured bubble, which at first may be considered as holes, but which in reality are enclosed by still thinner soap films. If bubbles like this are produced in a solid frame instead of in the hand, and if they are protected from evaporation in a vapour-filled enclosure, they can be preserved for days. Newton had disclosed that inside these black spots–*i.e.*, this membrane which weakly reflects light–still blacker, thus still thinner membranes are formed.

The thickness of the thinnest membrane obtained by Newton was about 6 millimicrons, about 20 times thinner than gold leaf. But the next thicker membrane obtainable, just before this thinnest one, is of exactly double thickness. This provided a tangible indication of the molecular structure of this skin substance; obviously, the thinnest membrane represents a single layer of molecules, while in the next thinnest one two layers are piled together. A very thin film can be produced more conveniently by allowing a minute amount of oil to spread out on a large water surface.

The familiar iridescent films which oil forms on water correspond in their thickness approximately to the usual soap bubble. Much thinner oil films which are no longer directly visible but can be recognized with simple tools can easily be formed. In these thicknesses of about 1 millimicron have been reached; and in these it has again been shown that the thicknesses of the finest films are not continuously variable. As in the soap bubbles, we find a quite definite thickness prescribed for the thinnest, second thinnest, etc. oil film (for a definite previously determined type of oil).

The limits of the divisibility of matter have actually been reached here. Further highly informative investigations were performed on very small corpuscular particles. Here the invention of the ultra-microscope rendered important service; it made visible not only the form, but also the position and motion of particles which cannot be "seen" in the usual sense, since there are limits, defined by the nature of light, for the optical observation of very small objects.

Bodies of dimensions smaller than the wavelength of visible light cannot be "formed" in any way by light rays; cannot be made visible in their true form. The ultra-microscope, through an ingenious device, made it possible to detect optically the presence of particles which are somewhat smaller than the wavelength of visible light.

For the fundamental task–to prove the reality of atoms experimentally–two types of investigations in particular have rendered real contributions. If very small particles are placed in a liquid it appears that they do not sink rectilinearly to the bottom and remain there but rather they move through the liquid in an irregular, erratic manner. The motion becomes more intense the higher the temperature of the liquid is raised; but it is independent of a minor agitation of the enclosing vessel or of other accidental disturbances. Also, if left undisturbed, the erratic motion of the particles continues for days or years.

This Brownian movement serves as a direct proof for the correctness of the proposition, set up by the atomic theory, that the heat content of a body is to be conceived of as an internal motion which is so fine that it cannot be macrophysically detected as such. One must not conclude that there are very fine motions or currents in the liquid which produce the Brownian movement of the suspended particles.

That would lead to the idea that the motions of a throng of closely neighboring particles were somewhat similar as is the case with dust we see around us in the sun's rays. In Brownian movements two approaching particles in a liquid are completely independent of each other, and thus separate again very soon. This shows that the hidden motions in the liquid, the permanent presence of which must be regarded as characteristic of the heat condition of the liquid, are completely "irregular" and macrophysically absolutely imperceptible.

Brownian movements have become the subject of many experimental investigations, for theoretical considerations had shown (Smoluchowski, Einstein) that its precise observation permits definite conclusions about the motions of atoms and molecules, through irregular collisions with which the particles in Brownian movements are driven around. It was possible to infer from investigation of Brownian movements how great the average kinetic energy of the individual molecules must be. And since the total heat energy contained in the liquid is equal to the sum of the energies of the individual molecules, we can infer how many molecules there are in a given volume. This leads to a determination of Loschmidt's number.

Related investigations and considerations are also feasible in many kinds of "fluctuation phenomena". Imagine an armor plate suspended from chains and blown upon by a very strong, uniformly operating sand blast apparatus; the plate will be forced from its equilibrium position by the pressure of the blast and will then hang in a slightly different position. Now, we remove the sand blast apparatus and by bombarding the plate with machine guns exert the same total pressure on the middle of the plate; the armor, struck over and over again by the single shots again assumes the displaced equilibrium position, this time continuously oscillating irregularly about this equilibrium position.

Observing just these oscillations, one can determine the strength of the single blows striking the plate. Similarly in many types of physical apparatus very fine irregular fluctuations can be observed and from them the magnitude of the Loschmidt number can be determined. In the very fine physical apparatus the presence of the irregular heat motion of the atoms can be detected; thereby most varied methods can be utilized for determining Loschmidt's number. The values of Loschmidt's number obtained from all such investigations have always agreed within the limits of their accuracy.

The following describes the second possible way of determining Loschmidt's number from investigations on very small particles: the density of the earth's atmosphere decreases, as we know, with increasing height. Imagine two very high cylinders–about 100 kilometers high–to be filled respectively with two different gases, perhaps oxygen and nitrogen. It is demonstrated that in the gas whose molecules are heavier (oxygen) the density decreases outward from the earth's surface more rapidly than in the other gas.

The gas molecules are hindered by their kinetic energy from simply accumulating on the bottom of the vessel, *i.e.*, forming a liquid. But the heavier they are, the stronger is the effect of gravity forcing them downwards and thus the more rapidly does the gas density decrease with increasing height above the earth's surface. The mathematical law relating to this states that the comparison of the decrease in density in the two cylinders depends solely on their molecular weights.

This was most useful in its application to the production of an artificial gas in which the actual mass of the individual molecules is known. Perrin laboriously produced large quantities of small resin pellets of equal weight. He obtained these resin particles from an emulsion which originally contained particles of the most varied magnitudes; but by an ingenious procedure, involving painstaking work, he was able to separate out particles of corresponding mass whose diameter was from 200 to 300 millimicrons.

These particles were then placed in water. (We know that with sufficient degree of dilution dissolved substances, *e.g.*, sugar in water, behave analogously to the ideal gases.) Thus, the resin particles placed in the water represented an ideal gas and vessels a kilometre high were no longer needed to determine the decrease in density with increasing height with this artificial gas; even in small containers these artificial gas molecules, enormously heavy in comparison with usual molecules, crowd together noticeably at the bottom. Through measurement of this effect and comparison with the known decrease of atmospheric density with increasing altitude it is possible to compare the directly determined mass of these artificial, huge molecules with the masses, absolute values still unknown, of the chemical molecules present in the atmosphere. Thus a new approach to the weighing of molecules–or differently expressed, to a determination of Loschmidt's number–is reached. The numerical value which resulted from these different determinations of Loschmidt's number (and from further determinations still to be discussed) is 6 x 1023 (a 6 with 23 zeros after it). If the hydrogen atoms contained in 100 grams of water were distributed over the entire earth's surface, one atom would fall on each square centimeter.. Finally we come to the experiments in which single atoms are actually isloated and made, in a manner of speaking, tangible, so that there is no longer any possible doubt concerning their real existence.

Let us begin with Laue's famous discovery of crystal interference. Modern workers in physical optics and spectroscopy no longer use only prisms for the spectral resolution of light; the diffraction grating has become much more important.

When numerous lines at equal, close intervals are scribed upon a polished surface, a monochromatic (including just one wave length) light ray will neither be reflected from this surface as from a usual mirror nor as from a rough surface which reflects diffusely towards all sides. The ray is reflected in a series of definite directions which vary according to its wave length; in all other directions there is no reflection because the light excitations traveling in these other directions mutually annul each other by interference.

This knowledge (a special interference experiment) gives us, as previously mentioned, the clearest and most obvious proof for the wave character of light and makes it possible to determine the wave length of the light just from, a knowledge of the spacing of the grating lines. (In common gratings this is between a hundredth and a thousandth of a millimeter.) Conversely naturally the separation of the rulings on the grating, if by chance this value is not known, can be determined from the reflection produced by this grating with light of known wave length.

X-rays, on traversing crystals, exhibit remarkable interference effects which are related to those of the "line grating" described but are considerably more complicated. The discovery of these effects proved two things with one stroke; first, that X-rays are a wave radiation; and second, that crystals possess a fine grating-like structure, which is so fine that it defied observation with former tools.

Careful analysis of the varied and complicated interference phenomena which can be obtained in this way led to the certainty that this internal fine structure of crystals is exactly the same as that which atomistic concepts led us to expect; th "illumination" of crystals with X-rays clearly displays their synthesis from atoms and thereby conclusively assures the reality of these atoms.

The wave lengths of X-rays are much shorter than those of ordinary light (X-rays, like light, are also included within the general bounds of the classification of Maxwell-Hertz electromagnetic waves). Because of their short wave lengths it is not possible without further refinement to attain diffraction (interference) of X-rays with the usual, relatively rough optical line gratings.

But in crystals nature presented us with "natural diffraction gratings" with which we can study X-ray interference effects. With a very refined device it was possible to obtain diffraction of X-rays with an optical diffraction grating (line grating) so that the wave lengths of X-rays could be measured just as optical wave lengths are. If we know the wave length of an X-ray, from the interference effects which result from the passage of this X-ray through a crystal we can infer the volume relations within the crystal. X-ray interference effects in crystals not only substantiate the reality of atoms with indubitable clearness and distinctness, but also facilitate another determination of the number of atoms in a macroscopic piece of matter–a new determination of Loschmidt's number. Millikan (in extending and refining older, less lucid ones) performed an experiment which was striking in the simplicity of its fundamental idea. In it he provided tangible proof for the atomic nature of electricity. A mist of minute oil drop lets is sprayed into a chamber; many of these are electrically charged in the process of their production.

One droplet is singled out and observed through a microscope. Left to itself it falls vertically downward, due to the force of gravity; not according to the ideal laws of free fall, but rather with constant velocity, since air resistance is proportionately more effective against these small droplets than against larger bodies. If the droplet is subjected to the influence of an electric field, of proper intensity perpendicular to its direction of fall and opposing it, the droplet will start to rise, with a constant velocity, the magnitude of which is dependent upon the field strength.

The ratio of the forces working in both cases and the magnitude of the charge carried by the droplet can be calculated from a comparison of the speeds of rising and falling. Consideration of the values obtained for numerous such droplets indicates that the magnitude of the charge does not vary continuously from case to case; none of them ever has a smaller charge than the so-called "elementary charge", and larger values are integral multiples (2, 3, 4...) of this smallest one. Thus we have tangible evidence that electric charges cannot be artibitrarily divided indefinitely; rather, all electric charges are composed of indivisible "elementary charges", each of which carries a charge of 4.77×10^{-10} (4.77 divided by a 1 with 10 zeros after it) electrostatic units.

This atomism of electricity is closely related to the atomism of matter in general. A law concerning electrolysis, discovered by Faraday and named after him, states this relation. Helmholtz had early concluded from this law that if the concept of the atomism of matter were proven correct an atomism of electricty must be assumed. If one zinc and one copper plate are immersed in an aqueous solution of copper sulfate and an electric current is passed through the solution in a positive direction from the zinc to the copper plate, the zinc gradually becomes dissolved in the liquid as further copper is deposited on the copper plate. Quantitatively this experiment shows that when one gram atomic weight of zinc has been dissolved, just one gram atomic weight of copper has been deposited; and corresponding weight relations are found to hold for all similar electrolytic experiments. Moreover, the number of gram atomic weights deposited or dissolved always remains in a simple proportion to the quantity of electric charge which has passed.

Thus there occur here regularities of the same character as the laws of weight relations in chemical reactions which Dalton explained by the concept of atomism. In these laws Dalton saw a criterion for the definition and use of the atomic concept; by the same right we can, with Helmholtz, infer the atomistic structure of electricity on the basis of Faraday's law.

Conversely, knowing from Millikan's work that the atomism of electricity not only expresses an auxiliary idea but is a real fact, we can infer from Faraday's law that the atomic structure of matter is a reality (which has already been verified in our consideration of crystal interference). The knowledge of the magnitude of the elementary electric charges furnishes one more basis for calculation of Loschmidt's number, characteristic of this atomistic structure of matter.

This in no way exhausts the experiments which show the indubitable reality of atoms. Further direct proofs for atomism are gained in connection with the investigation of radioactivity. From the observation of radioactive preparations it was determined that "alpha-radiation" is simply an emission of electrically charged helium. When this alpha-radiation strikes a suitably prepared zinc sulfide screen, even with limited intensity, it produces a series of small light flashes (scintillations) there.

From this effect it is seen that the emitted electrically charged helium is not an arbitrarily divisible, indefinitely diffuse substance, but must consist of discrete particles. By counting the light flashes, it became possible to determine how many single atoms constitute a definite quantity of helium; a new method of determining Loschmidt's number. The results again confirmed those obtained by other methods; this confirmation is evidence that each of these small light flashes actually does come from a single (electrically charged) helium atom.

Of even greater significance was still another process, in which again the effects of single atoms (or "ions", as electrically charged atoms are usually referred to) became visible. A moisture saturated atmosphere is produced in a carefully cleaned, dust-free vessel.

If the size of the container is suddenly increased by means of a moving piston the saturated atmosphere cools off and becomem "super saturated"–the water vapour begins to condense to liquid droplets. But the first droplets require "condensation nuclei" on which to form, and by removing all very fine particles (dust, etc.) we have eliminated the usual nuclei.

If an alpharadiation from a radioactive preparation strikes the atmosphere of this "Wilson Cloud Chamber", each of the rapidly traveling helium ions produces a fine track, along which are present innumerable atoms or molecules, now likewise electrically charged, which were hit by the alpha particles shooting through the air. Along the entire path these newly produced ions form suitable condensation nuclei; the entire path appears marked as a fine streak of mist visible to the naked eye. Again the effect of a single charged atom–and this time still more beautifully and definitely than by scintillations -has become visible, and again calculations from such "cloud tracks" yield Loschmidt's number.

The Wilson chamber has become one of the most important research tools of modern atomic physicists. Obviously its usefulness is limited to very rapidly flying particles, since only these possess sufficient energy to "ionise" innumerable other atoms or molecules along long paths. But in the processes connected with radioactivity these particles appear under various circumstances; and the Wilson chamber presents a beautiful opportunity to study all the details of the processes associated with their formation and transformation.

Geiger's development, the counting tube, is an equally important tool for atomic physics–for determining effects of single, isolated atomic particles. As soon as a single rapidly moving electrically charged atom enters this apparatus a macrophysically perceptible electric charge is produced by an ingenious arrangement. Like the Wilson chamber this apparatus makes it possible to establish the reality of atoms, to determine Loschmidt's number and to study thoroughly atomic physical processes.

Contributions of Atomic Physics

Before Millikan's determination of the elementary electric charge, electricity–notably through investigations by Lenard had been represented in pure culture, so to speak, namely in the form of cathode rays formed in a highly evacuated electric discharge tube–also in an X-ray tube, for example, where they produce X-rays by impinging on a solid plate. These cathode rays consist of a flow of negative electricity; from the results of Millikan's experiment it is certain that this electricity must flow in the form of separate, indivisibly small particles. These particles have been named "electrons"; they might also be designated as the atoms of electricity.

The cathode ray which travels rectilinearly when undisturbed can be forced into a curved path through the use of electric and magnetic fields. The motion of an electron deflected under the influence of known electric or magnetic fields must be calculable by Newton's second law–force equals mass times acceleration–if we know the ratio of its charge to its mass, since the acting force is proportional to the charge on a single electron.

Conversely, this relation of charge to mass can be inferred from the experimental determination of the deflection of cathode rays in electric and magnetic fields. Since the charge on an electron is known from Millikan's experiment (naturally from the outset it is assumed highly probably that the electrons in a cathode ray possess a single Millikan elementary charge, not two or three; further experiments confirm this as fact) the mass of an electron can be calculated from the measurements on cathode rays. The result shows that the mass of an electron is about 1838 times smaller than that of a hydrogen atom; or, expressed in grams, equals 0.9×10^{-27} (0.9 divided by a 1 with 27 zeros after it) grams.

The method of determining the ratio of charge to mass with cathode rays can also be extended to electrically charged atoms, ions. We know that from hydrogen atoms only one type of ion can be formed–a particle which has almost the same mass as the hydrogen atom itself and possesses exactly one positive elementary charge. Two different forms of ions–each of almost the same mass as the atom–can be formed from the helium atom; one has a single positive elementary charge, the other has two. The impression results that the electrically neutral hydrogen atom contains positive and negative charges within its structure. Obviously hydrogen contains just a single positive and a single negative electric charge; and the negative one must be connected with a proportionally minute mass, whereas almost the entire mass of the

Conversely, this relation of charge to mass can be inferred from the experimental determination of the deflection of cathode rays in electric and magnetic fields. Since the charge on an electron is known from Millikan's experiment (naturally from the outset it is assumed highly probably that the electrons in a cathode ray possess a single Millikan elementary charge, not two or three; further experiments confirm this as fact) the mass of an electron can be calculated from the measurements on cathode rays. The result shows that the mass of an electron is about 1838 times smaller than that of a hydrogen atom; or, expressed in grams, equals 0.9×10^{-27} (0.9 divided by a 1 with 27 zeros after it) grams.

The method of determining the ratio of charge to mass with cathode rays can also be extended to electrically charged atoms, ions. We know that from hydrogen atoms only one type of ion can be formed–a particle which has almost the same mass as the hydrogen atom itself and possesses exactly one positive elementary charge. Two different forms of ions–each of almost the same mass as the atom–can be formed from the helium atom; one has a single positive elementary charge, the other has two. The impression results that the electrically neutral hydrogen atom contains positive and negative charges within its structure.

Obviously hydrogen contains just a single positive and a single negative electric charge; and the negative one must be connected with a proportionally minute mass, whereas almost the entire mass of the hydrogen atom is attached to the positive one. The supposition is suggested–and abundant experience raises it to certainty–that the negative charge in the hydrogen atom is just a negative electron. The positive hydrogen ion, from which no further electrons can be separated, is designated as the "nucleus" of the hydrogen atom; it is also referred to as a "proton". In the helium atom two negative electrons are needed to neutralize the positive charge of the "nucleus"; here again ionization of the atom means simply the removal of one or both electrons. The nucleus of the helium atom is, moreover, identical with the aforementioned alpha particles.

As Rutherford recognized, the structure of all atoms is similar to these. In each atom almost the entire mass is concentrated in a positively charged nucleus; the effect of the "envelope" or "cloud" of negative electrons surrounding this nucleus is to render the atom electrically neutral. Thus the number of these electrons must always be equal to the "nuclear charge number"; *i.e.*, equal to the number of positive elementary charges in the heavy nucleus.

Chemists have known for a long time that a very convenient view of the chemical elements can be obtained through their arrangement in the "periodic table of the elements" -wherein the elements are arranged (Meyer and Mendeleef) according to their atomic weights. It was later shown that the atomic weight is not the characteristic by which the various chemical elements are differentiated that atoms of different atomic weights can belong to the same element (isotopes) and that atoms of equal atomic weights can belong to different elements (isobars). Not the mass of the nucleus, but its charge really determines the chemical nature of an atom; for the atomic number (nuclear charge number) determines the structure of the electron shells since it dictates how many electrons the neutral atom must possess. Through the structure of the electron shells the chemical properties are determined. Chemical molecules are formed when two or more atoms combine; in the combination the outer rings of electron shells experience certain transformations, but the nuclei remain unchanged.

The size of an atom can be determined in many ways. Their diameters are all equal to about 0.1 millicron. These diameters cannot be specified by exact numbers like the masses can, since the atoms are certainly not smooth spheres with definitely defined surfaces. As in a cloud with blurred boundaries, in an atom only an approximate value can be specified for its diameter. Such values can be determined from crystals, in which the atoms must be packed together very closely.

From investigations of gases it is possible to obtain corresponding determinations of these atomic magnitudes. The atomic nuclei are very much smaller; perhaps a hundred thousand times smaller in diameter than the atoms. Thus, almost the entire volume of an atom is occupied by the electron shells. But an electron is essentially no larger than an atomic nucleus. How it is possible that despite this, in a hydrogen atom for example where the entire "shell" consists of only one electron, the volume of the entire atom is "filled" up by this electron is a question to which we must return.

While chemical processes involve only the outer electron shells, there are other processes in which the nuclei themselves experience transformation, division or synthesis. In the main these are the already noted processes of radioactivity; but also numerous artificial, arbitrarily produced transmutations have been produced in atomic nuclei in recent years. Thus the nuclei must also be represented as complex structures; today we trace all nuclei back to two building stones, one of which is the already described proton and the other the "neutron"–a particle which has nearly the same mass as a proton but is electrically neutral.

Fundamental Forces

The fundamental forces are normally divided in four groups, of the four so-called "fundamental" forces. These are often naturally classified with respect to a dimensionless measure of their strength. To set these dimensions we use $\hbar$, c and the mass of the proton, m_p. The natural classification is then given in table.

Another important property is their range: the distance to which the interaction can be felt, and the type of quantity they couple to. Let me look a little closer at each of these in turn.

***Table:** A Summary of the Four Fundamental Forces*

Force	*Range*	*Strength*	*Acts on*
Gravity	∞	$G_N \approx 6\ 10^{-39}$	All particles (mass and energy)
Weak Force	$< 10^{-18}$m	$G_F \approx 1\ 10^{-5}$	Leptons, Hadrons
Electromagnetism	∞	1/137	All charged particles
Strong Force	$\approx 10^{-15}$m	$g^2 \approx 1$	Hadrons

In order to set the scale we need to express everything in a natural set of units. Three scales are provided by $\hbar$ and c and e-actually one usually works in units where these two quantities are 1 in high energy physics.

For the scale of mass we use the mass of the proton. In summary (for e = 1 we use electron volt as natural unit of energy):

$$\hbar = 6.58\times10^{-22} \text{ MeV s}$$

$$\hbar c = 1.97\times10^{-13} \text{ MeV}$$

$$m_p = 938 \text{ MeV}/c^2$$

Gravity

The theory of gravity can be looked at in two ways: The old fashioned Newtonian gravity, where the potential is proportional to the rest mass of the particles,

$$V = \frac{G_N m_1 m_2}{r}.$$

We find that $G_N m_p^2/\hbar c$ is dimensionless, and takes on the value,

$$G_N m_p^2/\hbar c = 5.9046486\times10^{-39}.$$

There are two more levels to look at gravity. One of those is Einstein's theory of gravity, which in the low-energy small-mass limit reduces to Newton's theory. This is still a *classical* theory, of a classical gravitational field.

The quantum theory, where we reexpress the field in their quanta has proven to be a very tough stumbling block-When one tries to generalise the approach taken for QED, every expression is infinite, and one needs to define an infinite number of different infinite constants. This is not deemed to be acceptable-*i.e.*, it doesn't define a theory. Such a model is called unrenormalisable. We may return to the problem of quantum gravity later, time permitting.

Electromagnetism

Electro-magnetism, *i.e.*, QED, has been discussed in some detail in the previous chapter.

Look there for a discussion. The coupling constant for the theory is,

$$\alpha = \frac{e^2}{4\pi\varepsilon_0 \hbar c}$$

Weak Force

This manifests itself through nuclear β decay, $\vec{n}p + e^- + \overline{\nu_e}$

The standard coupling for this theory is called the Fermi coupling, G_F, after its discoverer. After the theory was introduced it was learned that there were physical particles that mediate the weak force, the $W^{\pm}$ and the Z^0 bosons. These are very heavy particles (their mass is about 80 times the proton mass!), which is why they have such a small range-fluctuations where I need to create that much mass are rare. The $W^{\pm}$ bosons are charged, and the Z^0 boson is neutral. The typical β decay referred to above is mediated by a W^- boson as can be seen in the Feynman diagram figure.

The reason for this choice is that it conserves charge at each point (the charge of a proton and a W^- is zero, the charge of an electron and a neutrino is -1, the same as that of a W^-).

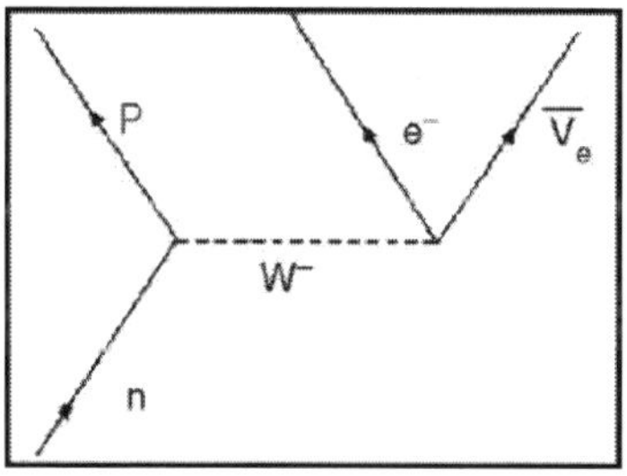

Figure; *The Feynman Diagram for the Weak Decay of a Neutron.*

Strong Force

The strong force is what keeps nuclei together. It is described by a theory called QCD, which described the forces between fermions called quarks that make up the hadrons. These forces are mediated by spin-1 bosons called gluons. Notice that this is a case where a series in powers of the coupling constant does not make a lot of sense, since higher powers have about the same value as lower powers. Such a theory is called non-perturbative.

Symmetries and Particle Physics

Symmetries in physics provide a great fascination to us-one of the hang-ups of mankind. We can recognise a symmetry easily, and they provide a great tool to classify shapes and patterns.

There is an important area of mathematics called group theory, where one studies the transformations under which an object is symmetric. In order to make this statement seem less abstract, let me look at a simple example, a regular hexagon in a plane. As can be seen in figure this object is symmetric (*i.e.*, we can't distinguish the new from the old object) under rotations around centre over angles of a multiple of 60°, and under reflection in any of the six axes sketched in the second part of the figure.

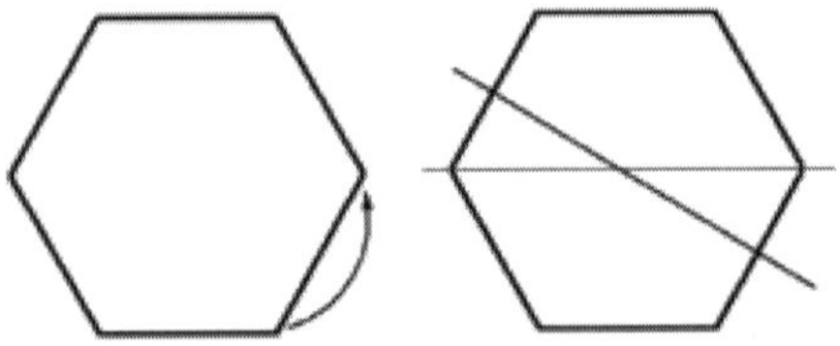

Figure: *The symmetries of a hexagon*

Importance of Symmetries: Noether's Theorem

There are important physical consequences of symmetries in physics, especially if the dynamics of a system is invariant under a symmetry transformation. There is a theorem, due to Emily Noether, one of the most important (female) mathematicians of this century, that states that for any *continuous* symmetry there is a conserved quantity. So what is a continuous symmetry?

Think about something like spherical symmetry-a sphere is invariant under any rotation about its centre, no matter what the rotation angle. The continuity of choice of parameter in a transformation is what makes the set of transformations continuous. Another way of saying the same thing is that the transformation can be arbitrarily close to the unit transformation, *i.e.*, it can do almost nothing at all.

Lorenz and Poincaré Invariance

One of the most common continuous symmetries of a relativistic theory is Lorentz invariance, *i.e.*, the dynamics is the same in any Lorentz frame. The group of Lorentz transformations can be decomposed into two parts:

- Boosts, where we go from one Lorentz frame to another, *i.e.*, we change the velocity.
- Rotations, where we change the orientation of the coordinate frame.

There is a slightly larger group of symmetries, called the Poincaré group, obtained when we add translations to the set of symmetries-clearly the dynamics doesn't care where we put the orbit of space. The set of conserved quantities associated with this group is large. Translational and boost invariance implies conservation of four momentum, and rotational invariance implies conservation of angular momentum.

Internal and Space-time Symmetries

Above I have mentioned angular momentum, the vector product of position and momentum. This is defined in terms of properties of space (or to be more generous, of space-time). But we know that many particles carry the spin of the particle to form the total angular momentum,

$$J = L + S.$$

The invariance of the dynamics is such that J is the conserved quantity, which means that we should not just rotate in ordinary space, but in the abstract "intrinsic space" where S is defined. This is something that will occur several times again, where a symmetry has a combination of a space-time and intrinsic part.

Discrete Symmetries

Let us first look at the key discrete symmetries-parity *P* (space inversion) charge conjugation *C* and time-reversal *T*.

Parity P

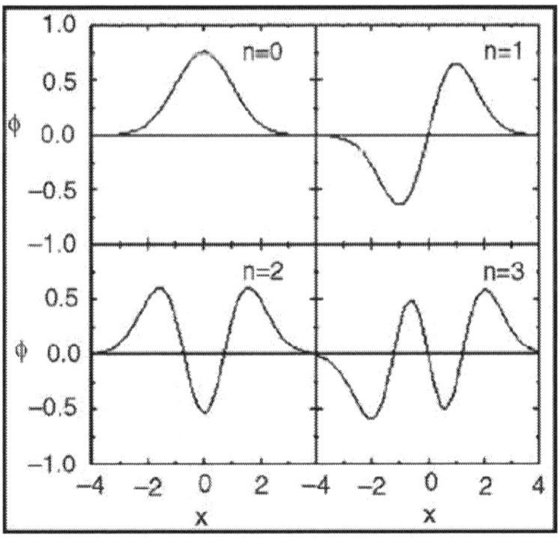

Figure: *The First Four Harmonic Oscillator Wave Functions*

Parity is the transformation where we reflect each point in the origin, $x \to -x$. This transformation should be familiar to you. Let us think of the one dimensional harmonic oscillator, with Hamiltonian,

$$\frac{\hbar^2}{2m}\frac{d}{dx^2}+\frac{1}{2}m\omega^2x^2.$$

The Hamiltonian does not change under the substitution $x \rightarrow -x$. The well-known eigenstates to this problem are either even or odd under this transformation, see Fig. and thus have either even or odd parity,

$$P\psi(x, t) = \psi(-x, t) = \pm\psi(x, t),$$

where P is the transformation that take $x \rightarrow -x$. For

$$P\psi(x, t) = \psi(x, t)$$

we say that the state has even parity, for the minus sign we speak about negative parity. These are the only two allowed eigenvalues, as can be seen from looking at the probability density $|\psi(x, y)|^2$. Since this must be *invariant*, we find that

$$|\psi(x, y)|^2 = |P\psi(x, y)|^2$$

which shows that the only *real* eigenvalues for P are ± 1. One can show that there is a relation between parity and the orbital angular momentum quantum number L, $\pi = (-1)^L$, which relates two space-time symmetries.

It is found, however, that parity also has an intrinsic part, which is associated with each type of particle. A photon (γ) has negative parity. This can be understood from the following classical analogy. When we look at Maxwell's equation for the electric field,

$$\nabla E(x,t) = \frac{1}{\varepsilon_0 \rho}(x,t),$$

we find that upon releversal of the coordinates this equation becomes,

$$\nabla E(-x,t) = \frac{1}{\varepsilon_0 \rho}(-x,t),$$

The additional minus sign, which originates in the change of sign of ∇ is what gives the electric field and thus the photon its negative intrinsic parity. We shall also wish to understand the parity of particles andantiparticles. Forfermions (electrons, protons...) we have the interesting relation $P_f P\bar{f} = -1$, which will come in handy later!

Charge conjugation C

The name of this symmetry is somewhat of a misnomer. Originally it stems from QED, where it was found that a set of interacting electrons behaves exactly the same way as a similar set of positrons. So if we change the sign of all charges the dynamics is the same. Actually, the symmetry generalises a little bit, and in general refers to a transformation where we change all particles in their antiparticles.

Once again we find $C^2 = 1$, and the only possible eigenvalues of this symmetry are ±1. An uncharged particle like the photon that is its own antiparticle, must be an eigenstate of the symmetry operation, and it is found that it has eigenvalue -1,

$$C\psi_\gamma = -\psi_\gamma.$$

(Here ψ_γ is the wave function of the photon.) This can be shown from Maxwell's equation as before, since ρ chanegs sign under charge conjugation. For a combination of a particle and an antiparticle, we find that $C_f C_{\overline{f}}$ =-1 for fermions, and +1 for bosons.

Time Reversal T

On a microscopic scale it is not very apparent whether time runs forward or backwards, the dynamics where we just change the sign of time is equally valid as the original one. This corresponds to flipping the sign of all momenta in a Feynman diagram, so that incoming particles become outgoing particles and vice-versa. This symmetry is slightly nastier, and acts on both space-time and intrinsic quantities such as spin in a complicated way. The space time part is found to be

$$T\ \psi(r,\ t) = \psi^{*}(r,-t).$$

Combined with its intrinsic part we find that it has eigenvalues $\pm i$ for fermions (electrons, etc.) and ±1 for bosons (photons, etc.).

The CPT Theorem

A little thought shows that all three symmetries mentioned above appear very natural-but that is a theorist's argument.

The real key test is experiment, not a theorist's nice ideas! In 1956 C.N. Yang and T.D. Lee analysed the experimental evidence for these symmetries. They realised there was good evidence of these symmetries in QED and QCD (the theory of strong interactions).

There was no evidence that parity was a symmetry of the weak interactions-which was true, since it was shown soon thereafter that these symmetries are broken, in a beautiful experiment led by "Madame" C.S. Wu.

There is a fairly strong proof that only minimal physical assumptions (locality, causality) that the product of *C*, *P* and *T* is a good symmetry of any theory. Up to now experiment has not shown any breaking of this product. We would have to rethink a lot of basic physics if this symmetry is not present. I am reasonably confident that if breaking is ever found there will be ten models that can describe it within a month!

CP Violation

The first experimental confirmation of symmetry breaking was found when studying β^- the decay of ^{60}Co,

$$^{60}Co \rightarrow {}^{60}Ni + e^- + \overline{\nu_e} \ .$$

This nucleus has a ground state with non-zero spin, which can be oriented in a magnetic field. A magnetic field is a pseudo-vector, which means that under parity it goes over into itself B $\rightarrow$ B. So does the spin of the nucleus, and we thus have established that under parity the situation under which the nucleus emits electrons should be invariant. But the direction in which they are emitted changes! Thus any asymmetry between the emission of electrons parallel and anti-parallel to the field implies parity breaking, as sketched in figure.

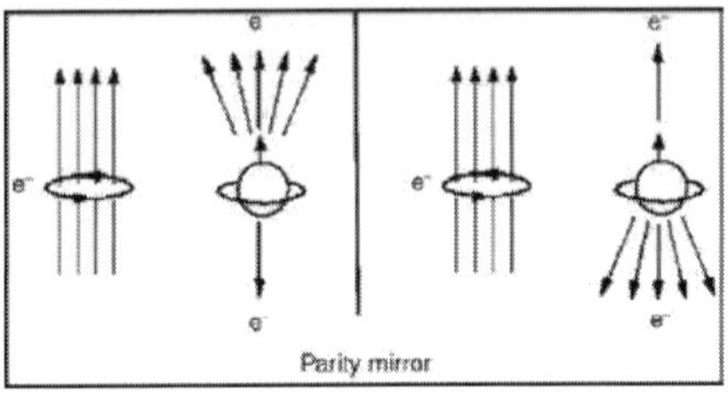

Figure: *Parity breaking for the* b *decay of* ^{60}Co

Actually one can shown that to high accuracy that the product of C and P is conserved, as can be seen in figure.

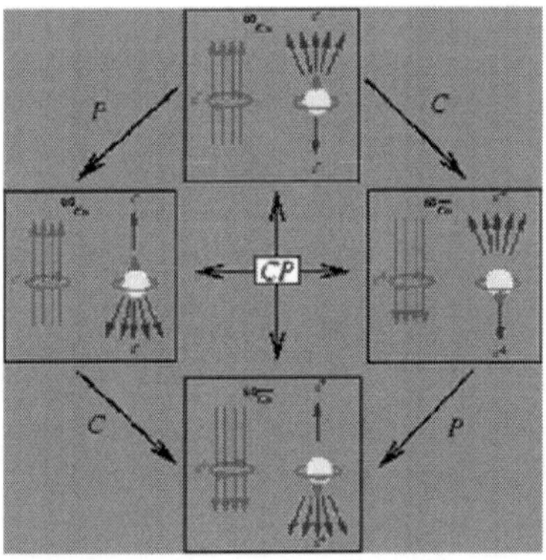

Figure: *CP symmetry for the decay of* ^{60}Co

Continuous Symmetries

Rotational Symmetry

Rotational transformations on a wave function can be applied by performing the transformation,

$$\exp\left[i\left(\theta_x \hat{J}_x + \theta_y \hat{J}_y + \theta_z \hat{J}_z\right)\right]$$

on a wave function. This is slightly simpler for a particle without spin, since we shall only have to consider the orbital angular momentum,

$$\hat{L} = \hat{p} \times r = i\hbar r \times \nabla.$$

Notice that this is still very complicated, exponentials of operators are not easy to deal with. One of the lessons we learn from applying this operator to many different states, is that if a state has good angular momentum *J*, the rotation can transform it into another state of angular momentum *J*, but it will never change the angular momentum.

This is most easily seen by labelling the states by J, M:

$$\left[\hat{J}_x^2 + \hat{J}_y^2 + \hat{J}_z^2\right]\phi JM = \hbar^2 J(J+1)\phi JM$$

$$\hat{J}_z^2 \phi JM = M\phi JM$$

The quantum number *M* can take the values-*J*,-*J* + 1..., *J*-1, *J*, so that we typically have 2*J* + 1 components for each *J*. The effect of the exponential transformation on a linear combination of states of identical *J* is to perform a linear transformation between these components. I shall show in a minute that such transformation can be implemented by unitary matrices. The transformations that implement these transformations are said to correspond to an irreducible representation of the rotation group (often denoted by SO(3)).

Let us look at the simplest example, for spin 1/2. We have two states, one with spin up and one with spin down $\psi_\pm$. If the initial state is $\psi = \alpha_+\psi_+ + \alpha-\psi-$, the effect of a rotation can only be to turn this into $\psi' = \alpha'_+\psi'_+ + \alpha'-\psi'-$. Since the transformation is linear (if I rotate the sum of two objects, I might as well rotate both of them) we find,

$$\begin{pmatrix} d_+ \\ d_- \end{pmatrix} = \begin{pmatrix} U_{++} & U_{-+} \\ U_{+-} & U_{--} \end{pmatrix} \begin{pmatrix} d_+ \\ d_- \end{pmatrix}$$

Since the transformation can not change the length of the vector, we must have, $\int|\psi'|^2 = 1$.

Assuming $\int|\psi_\pm|^2 = 1, \int\psi^\bullet + \psi - = 0$ we find $U^tU = 1$ with,

$$U^t = \begin{pmatrix} U^\bullet_{++} & U^\bullet_{+-} \\ U^\bullet_{-+} & U^\bullet_{--} \end{pmatrix}$$

the so-called hermitian conjugate.

We can write down matrices that in the space of $S = 1/2$ states behave the same as the angular momentum operators.

These are *half* the well known Pauli matrices,

$$\sigma_x = \begin{pmatrix} 0 & 1 \\ 1 & 0 \end{pmatrix}, \sigma_y = \begin{pmatrix} 0 & -i \\ i & 0 \end{pmatrix}, \sigma_z = \begin{pmatrix} 1 & 0 \\ 0 & -1 \end{pmatrix}$$

and thus we find that

$$U(\theta) = \exp\left[i(\theta x \sigma x/2 + \theta y \sigma y/2 + \theta z \sigma z/2)\right].$$

I don't really want to discuss how to evaluate the exponent of a matrix, apart from one special case. Suppose we perform a 2π rotation around the z axis, $\theta = (0, 0, 2\pi)$. We find,

$$U(0, 0, 2\pi) = \exp[i\pi \begin{pmatrix} 1 & 0 \\ 0 & -1 \end{pmatrix}].$$

Since this matrix is diagonal, we just have to evaluate the exponents for each of the entries (this corresponds to using the Taylor series of the exponential),

$$U(0, 0, 2\pi) = \begin{pmatrix} \exp[i\pi] & 0 \\ 0 & \exp[-i\pi] \end{pmatrix}$$

$$= \begin{pmatrix} -1 & 0 \\ 0 & -1 \end{pmatrix}.$$

To our surprise this does not take me back to where I started from. Let me make a small demonstration to show what this means. Finally what happens if we combine states from two irreducible representations? Let me analyse this for two spin 1/2 states,

$$\psi = (\alpha_+^1 \psi_+^1 + \alpha_-^1 \psi_-^1)(\alpha_+^2 \psi_+^2 + \alpha_-^2 \psi_-^2)$$

$$= \alpha_+^1 \alpha_+^2 \psi_+^1 \psi_+^2 + \alpha_+^1 \alpha_-^2 \psi_+^1 \psi_-^2 + \alpha_-^1 \alpha_+^2 \psi_-^1 \psi_+^2 \alpha_-^1 \alpha_-^2 \psi_-^1 \psi_-^2.$$

The first and the last product of ψ states have an angular momentum component ±1 in the z direction, and must does at least have $J = 1$. The middle two combinations with both have $M = M_1 + M_2 = 0$ can be shown to be a combination of a $J = 1$, $M = 0$ and a $J = 0$, $M = 0$ state. Specifically,

$$\frac{1}{\sqrt{2}}\left[\psi_+^1 \psi_-^2 \psi_-^1 \psi_+^2\right]$$

transforms as a scalar, it goes over into itself. the way to see that is to use the fact that these states transform with the same U, and substitute these matrices. The result is proportional to where we started from. Notice that the triplet (S = 1) is symmetric under interchange of the two particles, whereas the singlet (S = 0) is antisymmetric. This relation between symmetry can be exhibited where the horizontal direction denotes symmetry, and the vertical direction denotes antisymmetry. This technique works for all unitary groups.....

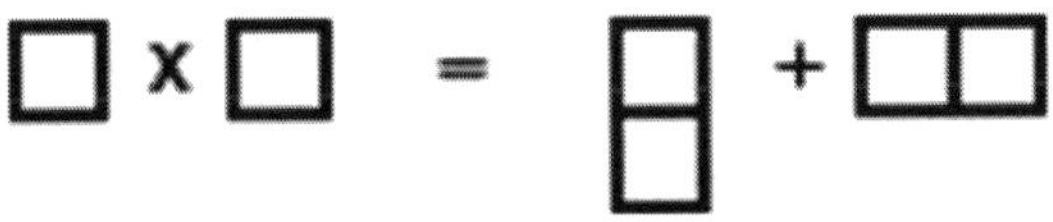

Figure: *The Young tableau for the multiplication 1/2×1/2 = 0 + 1.*

The coupling of angular momenta is normally performed through Clebsch-Gordan coefficients, as denoted by,

$$\langle j_1 m_1\, j_2 m_2 | JM \rangle.$$

We know that $M = m_1 + m_2$. Further analysis shows that J can take on all values $| j_1 - j_2 |, | j_1 - j_2 | + 1, | j_1 - j_2 | + 2, j1 + j2$.

Symmetries and Selection Rules

We shall often use the exact symmetries discussed up till now to determine what is and isn't allowed. Let us, for instance, look at

Representations of SU(3) and Multiplication Rules

A very important group is SU(3), since it is related to the colour carried by the quarks, the basic building blocks of QCD. The transformations within SU(3) are all those amongst a vector consisting of three complex objects that conserve the length of the vector. These are all three-by-three unitary matrices, which act on the complex vector ψ by,

$$\psi \to U\psi = \begin{pmatrix} U_{11} & U_{12} & U_{13} \\ U_{21} & U_{22} & U_{23} \\ U_{31} & U_{32} & U_{33} \end{pmatrix} \begin{pmatrix} \psi_1 \\ \psi_2 \\ \psi_3 \end{pmatrix}$$

The complex conjugate vector can be shown to transform as,

$$\psi^{\bullet} \to \psi^{\bullet} U^t,$$

with the inverse of the matrix. Clearly the fundamentalrepresentation of the group, where the matrices representing the transformation are just the matrix transformations, the vectors have length 3. The representation is usually labelled by its number of basis elements as 3. The one the transforms under the inverse matrices is usually denoted by $\overline{3}$.

What happens if we combine two of these objects, ψ and X˙? It is easy to see that the inner product of ψ and X˙ is scalar, X˙.$\psi \rightarrow$X˙ $UtU\psi$ = X˙.ψ where we have used the unitary properties of the matrices the remaining 8 components can all be shown to transform amongst themselves, and we write $3 \otimes \overline{3} = 1 \otimes 8$. Of further interest is the product of three of these vectors, $3 \otimes 3 \otimes 3 = 1 \otimes 8 \otimes 8 \otimes 10$.

Broken Symmetries

Of course one cannot propose a symmetry, discover that it is not realised in nature ("the symmetry is broken"), and expect that we learn something from that about the physics that is going on. But parity is broken, and we still find it a useful symmetry! That has to do with the manner in which it is broken, only weak interactions-the exchange of $W^{\pm}$ and Z bosons-break them. Any process mediated by strong, electromagnetic or (probably) gravitational forces conserves the symmetry. This is one example of a symmetry that is only mildly broken, *i.e.*, where the conserved quantities are still recognisable, even though they are not exactly conserved.

In modern particle physics the way symmetries are broken teaches us a lot about the underlying physics, and it is one of the goals of grand-unified theories (GUTs) to try and understand this.

Gauge Symmetries

One of the things which needs to be mentioned, is of a certain class of local symmetries (*i.e.*, symmetries of the theory at each point in space and time) called gauge symmetries. This is a key idea in almost all modern particle physics theories, so much so that they are usually labelled by the local symmetry group. Local symmetries are not directly observable, and do not have immediate consequences. They allow for a mathematically consistent and simple formulation of the theories, and in the end predict the particle that are exchanged-the gauge particles, as summarised in table.

Table: *The Four Fundamental Forces and their Gauge Particles*

Gravitation	Graviton(?)
QED	photon
Weak	$W^{\pm}$, Z^0
Strong	gluons

Symmetries of the Theory of Strong Interactions

The first time people realised the key role of symmetries was in the plethora of particles discovered using the firstaccelerators. Many of those were composite particle (to be explained later) bound by the strong interaction.

The First Symmetry: Isospin

The first particles that show an interesting symmetry are actually the nucleon and the proton. Their masses are remarkably close,

$$M_p = 939.566 \ MeV/c^2$$
$$M_n = 938.272 \ MeV/c^2.$$

If we assume that these masses are generated by the strong interaction there is more than a hint of symmetry here. Further indications come from the pions: they come in three charge states, and once again their masses are remarkably similar, M^{π^+}

$$= M^{\pi^-} = 139.567 \ MeV/c^2, \ M^{\pi^0} = 134.974 \ MeV/c^2.$$

This symmetry is reinforced by the discovery that the interactions between nucleon (p and n) is independent of charge, they only depend on the nucleon character of these particles-the strong interactions see only one nucleon and one pion. Clearly a continuous transformation between the nucleons and between the pions is a symmetry. The symmetry that was proposed (by Wigner) is an internal symmetry like spin symmetry called isotopic spin or isospin. It is an abstract rotation in isotopic space, and leads to similar type of states with isotopic spin I = 1/2, 1, 3/2.... One can define the third component of isospin as

$$Q = e(I_3 + B),$$

where B is the baryon number (B = 1 for n, p, 0 for π). We thus find,

	B	Q/e	I	I_3
n	1	0	1/2	-1/2
p	1	1	1/2	1/2
π^-	0	-1	1	-1
π^0	0	0	1	0
π^+	0	1	1	1

Notice that the energy levels of these particles are split by a magnetic force, as ordinary spins split under a magnetic force.

Strange Particles

The British physicists Rochester and Butler (from across the street) observed new particles in cosmic ray events. (Cosmic rays where the tool beforeaccelerators existed-they are still used due to the unbelievably violent processs taking place in the cosmos. We just can't produce particles like that in the lab. Unfortunately the number of highly energetic particles is very low, and we won't see many events.) These particles came in two forms: a neutral one that decayed into a π^+ and a π^-, and a positively charge one that decayed into a μ^+(heavy electron) and a photon, as sketched in figure.

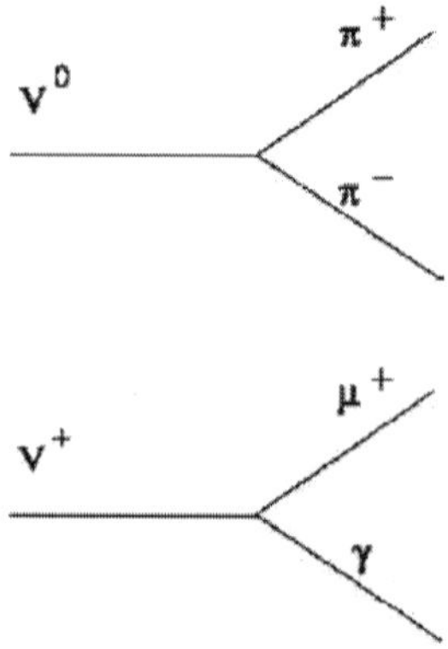

Figure: *The decay of* V *particles*

The big surprise about these particles was how long they lived. There are many decay time scales, but typically the decay times due to strong interactions are very fast, of the order of a femto second (10^{-15} s).

The decay time of the *K* mesons was about 10^{-10} s, much more typical of a weak decay. Many similar particles have since been found, both of mesonic and baryonic type (like pions or like nucleons).

These are collectively know as strange particles. Actually, using accelerators it was found that strange particles are typically formed in pairs, *e.g.*,

$$\pi^+ + \rightarrow \underbrace{\Lambda^\circ}_{\text{baryon}_+} \underbrace{K^\circ}\text{meson}$$

This mechanism was called associated production, and is highly suggestive of an additive conserved quantity, such as charge, called strangeness. If we assume that the has Λ^0 strangeness -1, and the K_0 +1, this balances

$$\pi^+ + p \rightarrow \Lambda^0 + K^0$$

$$0 + 0 -= -1 + 1$$

The weak decay

$$\Lambda^0 \rightarrow \pi^-$$

$$-1 \neq 0 + 0,$$

does not conserve strangeness (but it conserves baryon number). This process is indeed found to take much longer, about 10^{-10} s. Actually it is found (by analysing many resonance particles) that we can accommodate this quantity in our definition of isospin,

$$Q = e\left(I_3 + \frac{B+S}{2}\right)$$

Clearly for S =-1 and B = 1 we get a particle with $I3$ = 0. This allows us to identify the Λ^0 as an I = 0, $I_{3=0}$ particle, which agrees with the fact that there are no particles of different charge and a similar mass and strong interaction properties.

The kaons come in three charge states $K^{\pm}$, K^0 with masses $m_K{}^{\pm}$ = 494 *MeV*, $m_K{}^0$ = 498 *MeV*. In similarity with pions, which form an I = 1 multiplet, we would like to assume a I = 1 multiplet of *K*'s as well.

This is problematic since we have to assume S = 1 for all these particles: we cannot satisfy,

$$Q = e\left(I_3 + \frac{1}{2}\right)$$

for isospin 1 particles. The other possibility I = 3/2 doesn't fit with only three particles. Further analysis shows that the the K^+ is the antiparticle of K^-, but K^0 is not its own antiparticle (which is true for the pions. So we need four particles, and the assignments are S = 1, I = 1/2 for K^0 and K^-, S =-1, I = 1/2 for K^+ and $\overline{K}^0$. Actually, we now realise that we can summarise all the information about *K*'s and π's in one multiplet, suggestive of a (pretty badly broken!) symmetry.

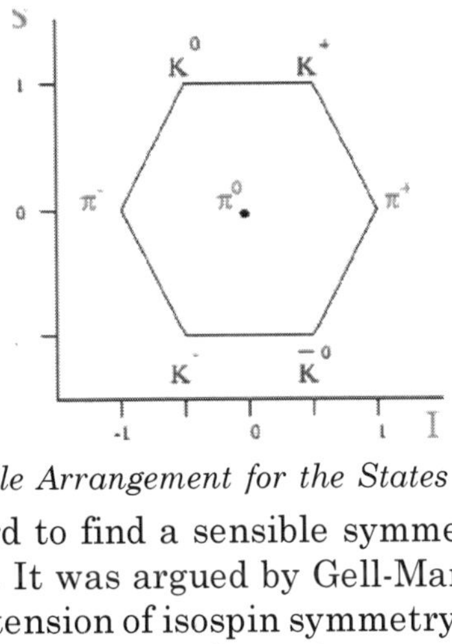

Figure: *A Possible Arrangement for the States of the Septet*

However, it is hard to find a sensible symmetry that gives a 7-dimensional multiplet. It was argued by Gell-Mann and Ne'eman in 1961 that a natural extension of isospin symmetry would be an SU(3) symmetry. We have argued before that one of the simplest representations of SU(3) is 8 dimensional symmetry. A mathematical analysis shows that what is missing is a particle with $I = I_3 = S = 0$. Such a particle is known, and is called the η^0. The breaking of the symmetry can be seen from the following mass table:

$$m_{\pi^\pm} = 139 \;MeV \qquad m_{\pi^0} = 134 \;MeV$$

$$m_{K^\pm} = 494 \;MeV \qquad m_{\overset{(-)}{k^0}} = 498 \;MeV$$

$$m_{\eta^0} = 549 \;MeV$$

The resulting multiplet is often represented like in figure.

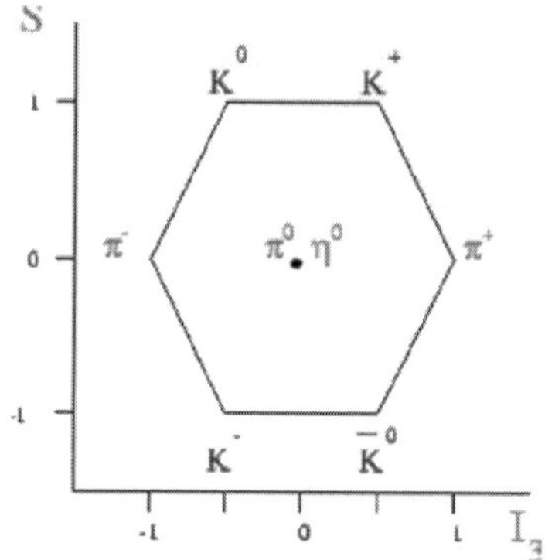

Figure: *Octet of mesons*

In order to have the scheme make sense we need to show its predictive power. This was done by studying the nucleons and their excited states. Since nucleons have baryon number one, they are labelled with the "hyper-charge" Y,

$$Y = (B + S),$$

rather than S. The nucleons form an octet with the single-strangeness particles Λ and σ and the doubly-strange cascade particle Ξ.

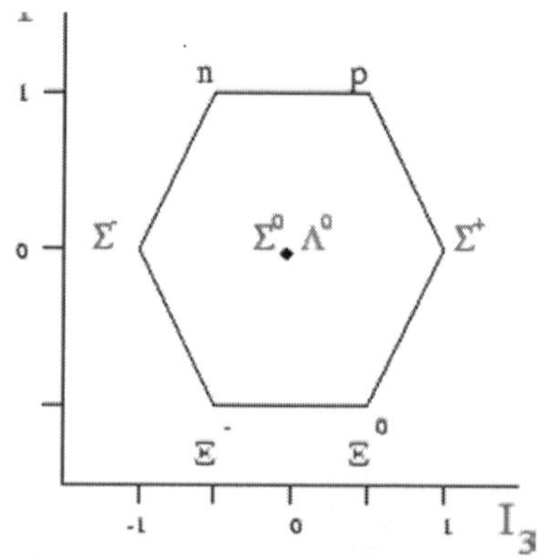

Figure: *Octet of nucleons*

The masses are:

$M_n = 938\ MeV$ $\quad M_p = 939\ MeV$

$M^{\Lambda^0} = 1115\ MeV$ $\quad M^{\Sigma^+} = 1189\ MeV$

$M^{\Sigma^0} = 1193\ MeV$ $\quad M^{\Sigma^-} = 1197\ MeV$

$M^{\Xi^0} = 1315\ MeV$ $\quad M^{\Xi^-} = 1321\ MeV$

All these particles were known before the idea of this symmetry. The first confirmation came when studying the excited states of the nucleon.

Nine states were easily incorporated in a decuplet, and the tenth state (the Ω^-, with strangeness -3) was predicted. It was found soon afterwards at the predicted value of the mass.

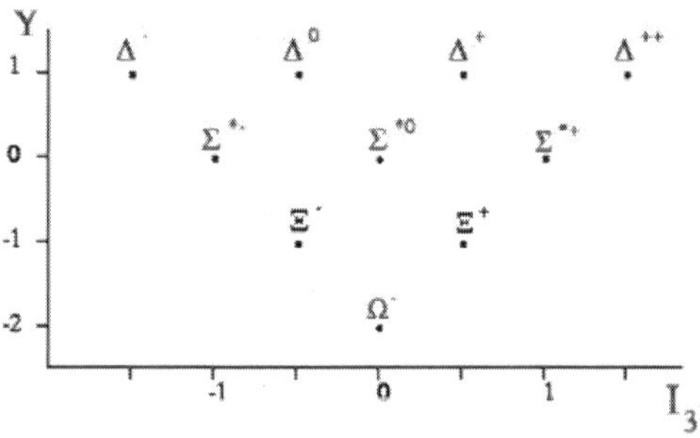

Figure: *decuplet of excited nucleons*

The masses are:

$M\Delta = 1232\ MeV$ $\quad M\Sigma^{\bullet} = 1385\ MeV$

$M\Xi^{\bullet} = 1530\ MeV$ $\quad M\Omega = 1672\ MeV$

Notice almost that we can fit these masses as a linear function in *Y*, as can be seen in figure. This was of great help in finding the Ω.

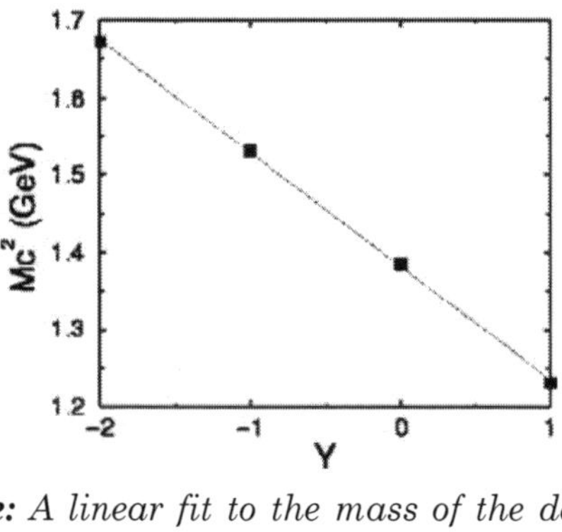

Figure: *A linear fit to the mass of the decuplet*

The Quark Model of Strong Interactions

Once the eightfold way (as the SU(3) symmetry was poetically referred to) was discovered, the race was on to explain it.

As I have shown before the decaplet and two octets occur in the product

$$3 \otimes 3 \otimes 3 = 1 \otimes 8 \otimes 8 \otimes 10.$$

A very natural assumption is to introduce a new particle that comes in three "flavours" called up, down and strange (u, d and s, respectively), and assume that the baryons are made from three of such particles, and the mesons from a quark and anti-quark (remember,

$$3 \otimes \bar{3} = 1 \otimes 8.)$$

Each of these quarks carries one third a unit of baryon number.

The properties can now be tabulated,

Table: *The properties of the three quarks.*

Quark	*label*	*spin*	*Q/e*	*I*	I_3	*S*	*B*
Up	u	$\frac{1}{2}$	$+\frac{2}{3}$	$\frac{1}{2}$	$+\frac{1}{2}$	0	$\frac{1}{3}$
Down	d	$\frac{1}{2}$	$-\frac{1}{3}$	$\frac{1}{2}$	$-\frac{1}{2}$	0	$\frac{1}{3}$
Strange	s	$\frac{1}{2}$	$-\frac{1}{3}$	0	0	-1	$\frac{1}{3}$

In the multiplet language I used before, we find that the quarks form a triangle, as given in Figure.

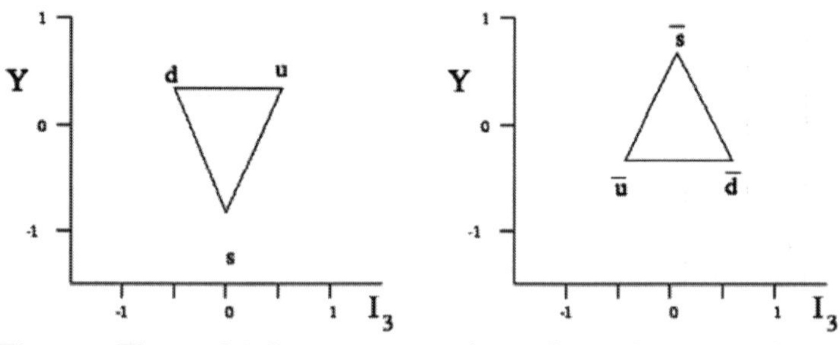

Figure: *The multiplet structure of quarks and antiquarks*

Once we have made this assigment, we can try to derive what combination corresponds to the assignments of the meson octet, figure.

We just make all possible combinations of a quark and ntiquark, apart from the scalar one $\eta' = u\bar{\mu} + d\bar{d} + c\bar{c}$ (why?).

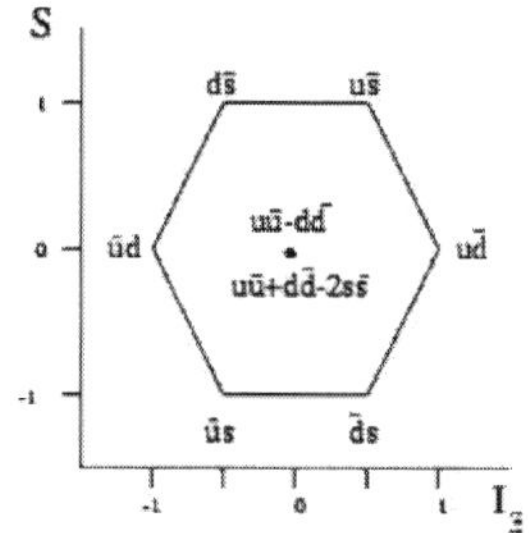

Figure: *Quark Assignment of the Meson octet*

A similar assignment can be made for the nucleon octet, and the nucleon decaplet,

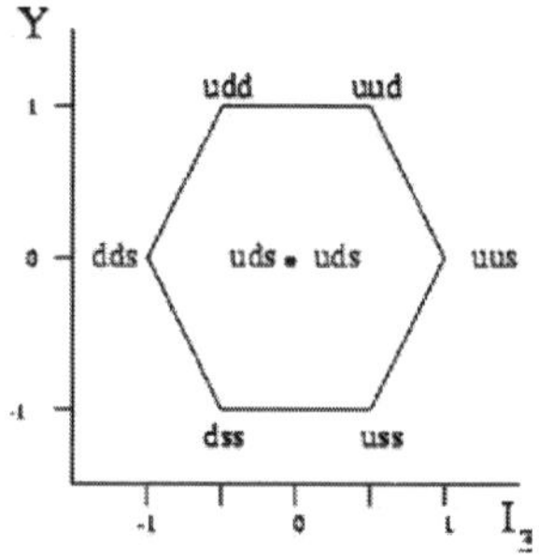

Figure: *quark assignment of the nucleon octet*

Once we have three flavours of quarks, we can ask the question whether more flavours exists. At the moment we know of three generations of quarks, corresponding to three generations (pairs). These give rise to SU(4), SU(5), SU(6) flavour symmetries. Since the quarks get heavier and heavier, the symmetries get more-and-more broken as we add flavours.

Table: *The properties of the three quarks.*

Quark	***label***	***spin***	***Q/e***	***mass (GEV/c²)***
Down	d	$\frac{1}{2}$	$-\frac{1}{3}$	0.35
Up	u	$\frac{1}{2}$	$+\frac{2}{3}$	0.35
Strange	s	$\frac{1}{2}$	$-\frac{1}{3}$	0.5
Charm	c	$\frac{1}{2}$	$+\frac{2}{3}$	1.5
Bottom	b	$\frac{1}{2}$	$-\frac{1}{3}$	4.5
Top	t	$\frac{1}{2}$	$+\frac{2}{3}$	93

Colour Symmetry

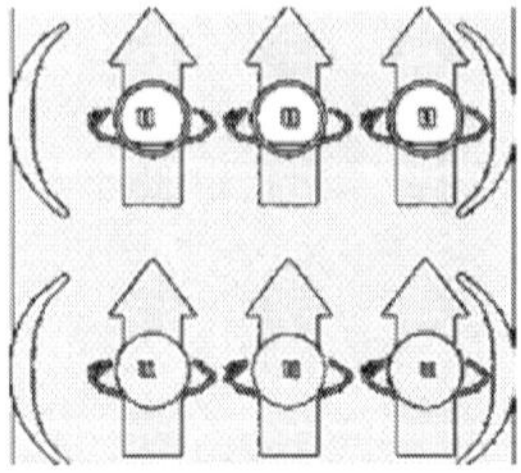

Figure: *The Δ^{++} in the quark model.*

So why don't we see fractional charges in nature? This is an important point! In so-called deep inelastic scattering we see pips inside the nucleon-these have been identified as the quarks. We do not see any direct signature of individual quarks. Furthermore, if quarks are fermions, as they are spin 1/2 particles, what about antisymmetry of their wavefunction? Let us investigate the Δ^{++}, see Fig. which consists of three u quarks with identical spin and flavour (isospin) and *symmetric* spatial wavefunction,

$$\psi_{total} = \psi_{space} \times \psi_{spin} \times \psi_{flavor},$$

This would be symmetric under interchange, which is unacceptable. Actually there is a simple solution. We "just" assume that there is an additional quantity called colour, and take the colour wave function to be antisymmetric:

$$\psi_{total} = \psi_{space} \times \psi_{spin} \times \psi_{flavor} \times \psi_{color}$$

We assume that quarks come in three colours. This naturally leads to yet another $SU(3)$ symmetry, which is actually related to the gauge symmetry of strong interactions, QCD. So we have shifted the question to: why can't we see coloured particles? This is a deep and very interesting problem. The only particles that have been seen are colour neutral ("white") ones. This leads to the assumption of confinement-We cannot liberate coloured particles at "low" energies and temperatures! The question whether they are free at higher energies is an interesting question, and is currently under experimental consideration.

The Feynman Diagrams of QCD

There are two key features that distinguish QCD from QED:

1. Quarks interact more strongly the further they are apart, and more weakly as they are close by-assymptotic freedom.
2. Gluons interact with themselves

The first point can only be found through detailed mathematical analysis. It means that free quarks can't be seen, but at high energies quarks look more and more like free particles. The second statement make QCD so hard to solve. The gluon comes in 8 colour combinations (since it carries a colour and anti-colour index, minus the scalar combination). The relevant diagrams are sketches in Figure. Try to work out yourself how we satisfy colour charge conservation!

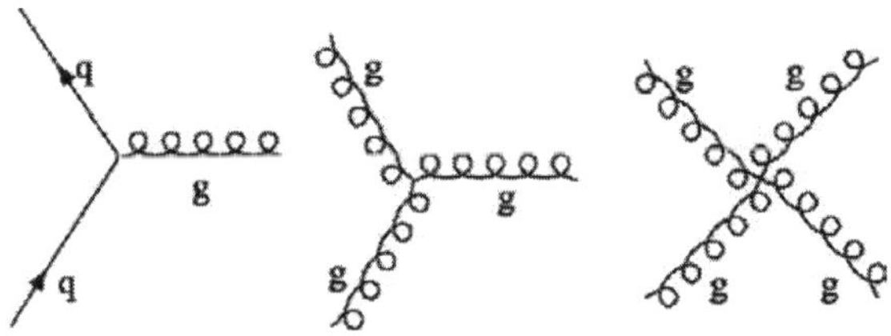

Figure: *The Basic Building Blocks for QCD Feynman Diagrams*

Jets and QCD

One way to see quarks is to use the fact that we can liberate quarks for a short time, at high energy scales. One such process is $e^+e^- \rightarrow q$, which use the fact that a photon can couple directly to $q\bar{q}$. The quarks don't live very long and decay by producing a "jet" a shower of particles that results from the deacay of the quarks. These are all "hadrons", mesons and baryons, since they must couple through the strong interaction.

By determining the energy in each if the two jets we can discover the energy of the initial quarks, and see whether QCD makes sense.

Chapter 4

Nuclear Properties

Radioactive Decay

Radioactive decay is the process by which an atomic nucleus of an unstable atom loses energy by emitting ionizing particles (ionizing radiation). The emission is spontaneous, in that the atom decays without any interaction with another particle from outside the atom (i.e., without a nuclear reaction). Usually, radioactive decay happens due to a process confined to the nucleus of the unstable atom, but, on occasion (as with the different processes of electron capture and internal conversion), an inner electron of the radioactive atom is also necessary to the process.

Radioactive decay is a stochastic (i.e., random) process at the level of single atoms, in that, according to quantum theory, it is impossible to predict when a given atom will decay. However, given a large number of identical atoms (nuclides), the decay rate for the collection is predictable, via the Law of Large Numbers. The decay, or loss of energy, results when an atom with one type of nucleus, called the *parent radionuclide*, transforms to an atom with a nucleus in a different state, or a different nucleus, either of which is named the *daughter nuclide.* Often th parent and daughter are different chemical elements, and in such cases the decay process results in nuclear transmutation. In an example of this, a carbon-14 atom (the "parent") emits radiation (a beta particle, antineutrino, and a gamma ray) and transforms to a nitrogen-14 atom (the "daughter").

By contrast, there exist two types of radioactive decay processes (gamma decay and internal conversion decay) that do not result in transmutation, but only decrease the energy of an excited nucleus.

This results in an atom of the same element as before but with a nucleus in a lower energy state. An example is the nuclear isomer technetium-99m decaying, by the emission of a gamma ray, to an atom of technetium-99. Nuclides produced as daughters are called radiogenic nuclides, whether they themselves are stable or not. A number of naturally occurring radionuclides are short-lived radiogenic nuclides that are the daughters of radioactive primordial nuclides (types of radioactive atoms that have been present since the beginning of the Earth and solar system).

Other naturally occurring radioactive nuclides are cosmogenic nuclides, formed by cosmic ray bombardment of material in the Earth's atmosphere or crust. For a summary table showing the number of stable nuclides and of radioactive nuclides in each category, see Radionuclide. The SI unit of activity is the becquerel (Bq). One Bq is defined as one transformation (or decay) per second. Since any reasonably-sized sample of radioactive material contains many atoms, a Bq is a tiny masure of activity; amounts on the order of GBq (gigabecquerel, 1×10^9 decays per second) or TBq (terabecquerel, 1×10^{12} decays per second) are commonly used. Another unit of radioactivity is the curie, Ci, which was originally defined as the amount of radium emanation (radon-222) in equilibrium with one gram of pure radium, isotope Ra-226. At present it is equal, by definition, to the activity of any radionuclide decaying with a disintegration rate of 3.7×10^{10} Bq. The use of Ci is presently discouraged by the SI.

Explanation

The neutrons and protons that constitute nuclei, as well as other particles that approach close enough to them, are governed by several interactions. The strong nuclear force, not observed at the familiar macroscopic scale, is the most powerful force over subatomic distances. The electrostatic force is almost always significant, and, in the case of beta decay, the weak nuclear force is also involved. The interplay of these forces produces a number of different phenomena in which energy may be released by rearrangement of particles in the nucleus or the change of one particle into others.

The rearrangement is hindered energetically, so that it does not occur immediately. Random quantum vacuum fluctuations are theorized to promote relaxation to a lower energy state (the "decay") in a phenomenon known as quantum tunneling. One might draw an analogy with a snowfield on a mountain: While friction between the ice crystals may be supporting the snow's weight, the system is inherently unstable with regard to a state of lower potential energy. A disturbance would thus facilitate the path to a state of greater entropy:

The system will move towards the ground state, producing heat, and the total energy will be distributable over a larger number of quantum states. Thus, an avalanche results. The total energy does not change in this process, but, because of the law of entropy, avalanches happen only in one direction and that is toward the "ground state" — the state with the largest number of ways in which the available energy could be distributed.

Such a collapse (a *decay event*) requires a specific activation energy. For a snow avalanche, this energy comes as a disturbance from outside the system, although such disturbances can be arbitrarily small. In the case of an excited atomic nucleus, the arbitrarily small disturbance comes from quantum vacuum fluctuations. A radioactive nucleus (or any excited system in quantum mechanics) is unstable, and can, thus, *spontaneously* stabilize to a less-excited system. The resulting transformation alters the structure of the nucleus and results in the emission of either a photon or a high-velocity particle that has mass (such as an electron, alpha particle, or other type).

Discovery

Radioactivity was first discovered in 1896 by the French scientist Henri Becquerel, while working on phosphorescent materials. These materials glow in the dark after exposure to light, and he suspected that the glow produced in cathode ray tubes by X-rays might be connected with phosphorescence. He wrapped a photographic plate in black paper and placed various phosphorescent salts on it. All results were negative until he used uranium salts. The result with these compounds was a deep blackening of the plate. These radiations were called Becquerel Rays. It soon became clear that the blackening of the plate had nothing to do with phosphorescence, because the plate blackened when the mineral was in the dark. Non-phosphorescent salts of uranium and metallic uranium also blackened the plate. It was clear that there is a form of radiation that could pass through paper that was causing the plate to become black. At first it seemed that the new radiation was similar to the then recently-discovered X-rays. Further research by Becquerel, Ernest Rutherford, Paul Villard, Marie Curie, Pierre Curie, and others discovered that this form of radioactivity was significantly more complicated. Different types of decay can occur, producing very different types of radiation. Rutherford was the firstto realize that they all occur with the same mathematical exponential formula, and to realize that many decay processes resulted in the transmutation of one element to another. The early researchers also discovered that many other chemical elements besides uranium have radioactive isotopes. A systematic search for the total radioactivity in uranium ores also guided Marie Curie to isolate a new element polonium and to separate a new element radium from barium. The two elements' chemical similarity would otherwise have made them difficult to distinguish.

Danger of Radioactive Substances

The dangers of radioactivity and radiation were not immediately recognized. Acute effects of radiation were first observed in the use of X-rays when electrical engineer and physicist Nikola Tesla intentionally subjected his fingers to X-rays in 1896. He published his observations concerning the burns that developed, though he attributed them to ozone rather than to X-rays. His injuries healed later. The genetic effects of radiation, including the effects on cancer risk, were recognized much later.

In 1927, Hermann Joseph Muller published research showing genetic effects, and in 1946 was awarded the Nobel prize for his findings. Before the biological effects of radiation were known, many physicians and corporations had begun marketing radioactive substances as patent medicine, glow-in-the-dark pigments, and radioactive quackery. Examples were radium enema treatments, and radium-containing waters to be drunk a tonics. Marie Curie spoke out against this sort of treatment, warning that the effects of radiation on the human body were not well understood (Curie later died from aplastic anemia, which was likely caused by exposure to ionizing radiation). By the 1930s, after a number of cases of bone necrosis and death in enthusiasts, radium-containing medical products had nearly vanished from the market.

Types of Decay

As for types of radioactive radiation, it was found that an electric or magnetic field could split such emissions intothree types of beams. For lack of better terms, the rays were given the alphabetic names alpha, beta, and gamma, still in use today. While alpha decay was seen only in heavier elements (atomic number 52, tellurium, and greater), the other two types of decay were seen in all of the elements. In analyzing the nature of the decay products, it was obvious from the direction of electromagnetic forces produced upon the radiations by external magnetic and electric fields that alpha rays carried a positive charge, beta rays carried a negative charge, and gamma rays were neutral.

From the magnitude of deflection, it was clear that alpha particles were much more massive than beta particles. Passing alpha particles through a very thin glass window and trapping them in a discharge tube allowed researchers to study the emission spectrum of the resulting gas, and ultimately prove that alpha particles are helium nuclei. Other experiments showed the similarity between classical beta radiation and cathode rays: They are both streams of electrons. Likewise gamma radiation and X-rays were found to be similar high-energy electromagnetic radiation. The relationship between types of decays also began to be examined: For example, gamma decay was almost always found associated with other types of decay, occurring at about the same time, or afterward. Gamma decay as a separate phenomenon (with its own half-life, now termed isomeric transition), was found in natural radioactivity to be a result of the gamma decay of excited metastable nuclear isomers, in turn created from other types of decay.

Although alpha, beta, and gamma were found most commonly, other types of decay were eventually discovered. Shortly after the discovery of the positron in cosmic ray products, it was realized that the same process that operates in classical beta decay can also produce positrons (positron emission).

In an analogous process, instead of emitting positrons and neutrinos, some proton-rich nuclides were found to capture their own atomic electrons (electron capture), and emit only a neutrino (and usually also a gamma ray). Each of these types of decay involves the capture or emission of nuclear electrons or positrons, and acts to move a nucleus toward the ratio of neutrons to protons that has the least energy for a given total number of nucleons (neutrons plus protons). Shortly after discovery of the neutron in 1932, it was discovered by Enrico Fermi that certain rare decay reactions yield neutrons as a decay particle (neutron emission). Isolated proton emission was eventually observed in some elements. It was also found that some heavy elements may undergo spontaneous fission into products that vary in composition. In a phenomenon called cluster decay, *specific combinations of neutrons and protons* (atomic nuclei) other than alpha particles (helium nuclei) were found to be spontaneously emitted from atoms, on occasion. Other types of radioactive decay that emit previously seen particles were found, but by different mechanisms. An example is internal conversion, which results in electron and sometimes high-energy photon emission, even though it involves neither beta nor gamma decay. This type of decay (like isomeric transition gamma decay) did not transmute one element to another. Rare events that involve a combination of two beta-decay type events happening simultaneously are known. Any decay process that does not violate conservation of energy or momentum laws (and perhaps other particle conservation laws) is permitted to happen, although not all have been detected. An interesting example (discussed in a final section) is bound state beta decay of rhenium-187. In this process, an inverse of electron capture, beta electron-decay of the parent nuclide is not accompanied by beta electron emission, because the beta particle has been captured into the K-shell of the emitting atom. An antineutrino, however, is emitted.

Decay Modes in Table Form

Radionuclides can undergo a number of different reactions. These are summarized in the following table. A nucleus with mass number A and atomic number Z is represented as (A, Z). The column "Daughter nucleus" indicates the difference between the new nucleus and the original nucleus. Radioactive decay results in a reduction of summed rest mass, once the released energy (the *disintegration energy*) has escaped in some way (for example, the products might be captured and cooled, and the heat allowed to escape).

Although decay energy is sometimes defined as associated with the difference between the mass of the parent nuclide products and the mass of the decay products, this is true only of rest mass measurements, where some energy has been removed from the product system. This is true because the decay energy must always carry mass with it, wherever it appears according to the formula $E = mc^2$. The decay energy is initially released as the energy of emitted photons plus the kinetic energy of massive emitted particles (that is, particles that have rest mass).

If these particles come to thermal equilibrium with their surroundings and photons are absorbed, then the decay energy is transformed to thermal energy, which retains its mass. Decay energy therefore remains associated with a certain measure of mass of the decay system invariant mass. The energy of photons, kinetic energy of emitted particles, and, later, the thermal energy of the surrounding matter, all contribute to calculations of invariant mass of systems. Thus, while the sum of rest masses of particles is not conserved in radioactive decay, the *system* mass and system invariant mass (and also the system total energy) is conserved throughout any decay process.

Decay Chains and Multiple Modes

The daughter nuclide of a decay event may also be unstable (radioactive). In this case, it will also decay, producing radiation. The resulting second daughter nuclide may also be radioactive. This can lead to a sequence of several decay events. Eventually, a stable nuclide is produced. This is called a *decay chain*. An example is the natural decay chain of ^{238}U which is as follows:

- decays, through alpha-emission, with a half-life of 4.5 billion years to thorium-234
- which decays, through beta-emission, with a half-life of 24 days to protactinium-234
- which decays, through beta-emission, with a half-life of 1.2 minutes to uranium-234
- which decays, through alpha-emission, with a half-life of 240 thousand years to thorium-230
- which decays, through alpha-emission, with a half-life of 77 thousand years to radium-226
- which decays, through alpha-emission, with a half-life of 1.6 thousand years to radon-222
- which decays, through alpha-emission, with a half-life of 3.8 days to polonium-218
- which decays, through alpha-emission, with a half-life of 3.1 minutes to lead-214
- which decays, through beta-emission, with a half-life of 27 minutes to bismuth-214
- which decays, through beta-emission, with a half-life of 20 minutes to polonium-214
- which decays, through alpha-emission, with a half-life of 160 microseconds to lead-210
- which decays, through beta-emission, with a half-life of 22 years to bismuth-210
- which decays, through beta-emission, with a half-life of 5 days to polonium-210
- which decays, through alpha-emission, with a half-life of 140 days to lead-206, which is a stable nuclide.

Some radionuclides may have several different paths of decay. For example, approximately 36% of bismuth-212 decays, through alpha-emission, to thallium-208 while approximately 64% of bismuth-212 decays, through beta-emission, to polonium-212. Both the thallium-208 and the polonium-212 are radioactive daughter products of bismuth-212, and both decay directly to stable lead-208.

Occurrence and Applications

According to the Big Bang theory, stable isotopes of the lightest five elements (H, He, and traces of Li, Be, and B) were produced very shortly after the emergence of the universe, in a process called Big Bang nucleosynthesis. These lightest stable nuclides (including deuterium) survive to today, but any radioactive isotopes of the light elements produced in the Big Bang (such as tritium) have long since decayed. Isotopes of elements heavier than boron were not produced at all in the Big Bang, and these first five elements do not have any long-lived radioisotopes. Thus, all radioactive nuclei are, therefore, relatively young with respect to the birth of the universe, having formed later in various other types of nucleosynthesis in stars (in particular, supernovae), and also during ongoing interactions between stable isotopes and energetic particles. For example, carbon-14, a radioactive nuclide with a half-life of only 5730 years, is constantly produced in Earth's upper atmosphere due to interactions between cosmic rays and nitrogen. Radioactive decay has been put to use in the technique of radioisotopic labeling, which is used to track the passage of a chemical substance through a complex system (such as a living organism).

A sample of the substance is synthesized with a high concentration of unstable atoms. The presence of the substance in one or another part of the system is determined by detecting the locations of decay events. On the premise that radioactive decay is truly random (rather than merely chaotic), it has been used in hardware random-number generators. Because the process is not thought to vary significantly in mechanism over time, it is also a valuable tool in estimating the absolute ages of certain materials. For geological materials, the radioisotopes and some of their decay products become trapped when a rock solidifies, and can then later be used (subject to many well-known qualifications) to estimate the date of the solidification. These include checking the results of several simultaneous processes and their products against each other, within the same sample.

In a similar fashion, and also subject to qualification, the rate of formation of carbon-14 in various eras, the date of formation of organic matter within a certain period related to the isotope's half-life may be estimated, because the carbon-14 becomes trapped when the organic matter grows and incorporates the new carbon-14 from the air. Thereafter, the amount of carbon-14 in organic matter decreases according to decay processes that may also be independently cross-checked by other means (such as checking the carbon-14 in individual tree rings, for example). An example is the natural decay chain of ^{238}U which is as follows:

- decays, through alpha-emission, with a half-life of 4.5 billion years to thorium-234
- which decays, through beta-emission, with a half-life of 24 days to protactinium-234
- which decays, through beta-emission, with a half-life of 1.2 minutes to uranium-234
- which decays, through alpha-emission, with a half-life of 240 thousand years to thorium-230
- which decays, through alpha-emission, with a half-life of 77 thousand years to radium-226
- which decays, through alpha-emission, with a half-life of 1.6 thousand years to radon-222
- which decays, through alpha-emission, with a half-life of 3.8 days to polonium-218
- which decays, through alpha-emission, with a half-life of 3.1 minutes to lead-214
- which decays, through beta-emission, with a half-life of 27 minutes to bismuth-214
- which decays, through beta-emission, with a half-life of 20 minutes to polonium-214
- which decays, through alpha-emission, with a half-life of 160 microseconds to lead-210
- which decays, through beta-emission, with a half-life of 22 years to bismuth-210
- which decays, through beta-emission, with a half-life of 5 days to polonium-210
- which decays, through alpha-emission, with a half-life of 140 days to lead-206, which is a stable nuclide.

Some radionuclides may have several different paths of decay. For example, approximately 36% of bismuth-212 decays, through alpha-emission, to thallium-208 while approximately 64% of bismuth-212 decays, through beta-emission, to polonium-212. Both the thallium-208 and the polonium-212 are radioactive daughter products of bismuth-212, and both decay directly to stable lead-208.

Occurrence and Applications

According to the Big Bang theory, stable isotopes of the lightest five elements (H, He, and traces of Li, Be, and B) were produced very shortly after the emergence of the universe, in a process called Big

Bang nucleosynthesis. These lightest stable nuclides (including deuterium) survive to today, but any radioactive isotopes of the light elements produced in the Big Bang (such as tritium) have long since decayed. Isotopes of elements heavier than boron were not produced at all in the Big Bang, and these first five elements do not have any long-lived radioisotopes.

Thus, all radioactive nuclei are, therefore, relatively young with respect to the birth of the universe, having formed later in various other types of nucleosynthesis in stars (in particular, supernovae), and also during ongoing interactions between stable isotopes and energetic particles. For example, carbon-14, a radioactive nuclide with a half-life of only 5730 years, is constantly produced in Earth's upper atmosphere due to interactions between cosmic rays and nitrogen. Radioactive decay has been put to use in the technique of radioisotopic labeling, which is used to track the passage of a chemical substance through a complex system (such as a living organism). A sample of the substance is synthesized with a high concentration of unstable atoms. The presence of the substance in one or another part of the system is determined by detecting the locations of decay events.

On the premise that radioactive decay is truly random (rather than merely chaotic), it has been used in hardware random-number generators. Because the process is not thought to vary significantly in mechanism over time, it is also a valuable tool in estimating the absolute ages of certain materials. For geological materials, the radioisotopes and some of their decay products become trapped when a rock solidifies, and can then later be used (subject to many well-known qualifications) to estimate the date of the solidification. These include checking the results of several simultaneous processes and their products against each other, within the same sample. In a similar fashion, and also subject to qualification, the rate of formtion of carbon-14 in various eras, the date of formation of organic matter within a certain period related to the isotope's half-life may be estimated, because the carbon-14 becomes trapped when the organic matter grows and incorporates the new carbon-14 from the air. Thereafter, the amount of carbon-14 in organic matter decreases according to decay processes that may also be independently cross-checked by other means (such as checking the carbon-14 in individual tree rings, for example).

Radioactive Decay Rates

The decay rate, or activity, of a radioactive substance are characterized by:

Constant quantities:

- half life — symbol $t_{1/2}$ — the time taken for the activity of a given amount of a radioactive substance to decay to half of its initial value.
- mean lifetime — symbol τ — the average lifetime of a radioactive particle.
- decay constant — symbol $\ddot{e}$ — the inverse of the mean lifetime.

Although these are constants, they are associated with statistically random behaviour of populations of atoms. In consequence predictions using these constants are less accurate for small number of atoms.

Changing Decay Rates

The radioactive decay modes of electron capture and internal conversion are known to be slightly sensitive to chemical and environmental effects which change the electronic structure of the atom, which in turn affects the presence of 1s and 2s electrons that participate in the decay process. A small number of mostly light nuclides are affected. For example, chemical bonds can affect the rate of electron capture to a small degree (in general, less than 1%) depending on the proximity of electrons to the nucleus in beryllium. In ^{7}Be, a difference of 0.9% has been observed between half-lives in metallic and insulating environments. This relatively large effect is because beryllium is a small atom whose valence electrons are in 2s atomic orbitals, which have a large degree of penetration very close to the nucleus, and thus are subject to electron capture. Rhenium-187 is a more spectacular example. ^{187}Re normally beta decays to ^{187}Os with a half life of 41.6×10^{9} y, but studies using fully ionised ^{187}Re atoms (bare nuclei) have found that this can decrease to only 33 y. This is attributed to "bound-state β^{-} decay" of the fully ionised atom — the electron is emitted into the K-shell (1s orbital), which cannot occur for neutral atoms in which all low-lying bound states are occupied.

A number of experiments have found that decay rates of other modes of artificial and naturally-occurring radioisotopes are, to a high degree of precision, unaffected by external conditions such as temperature, pressure, the chemical environment, and electric, magnetic, or gravitational fields.

Comparison of laboratory experiments over the last century, studies of the Oklo natural nuclear reactor (which exemplified the effects of thermal neutrons on nuclear decay), and astrophysical observations of the luminosity decays of distant supernovae (which occurred far away so the light has taken a great deal of time to reach us), for example, strongly indicate that decay rates have been constant (at least to within the limitations of small experimental errors) as a function of time as well. On the other hand, some recent results suggest the possibility that decay rates might have a weak dependence (0.5% or less) on environmental factors.

It has been suggested that measurements of decay rates of silicon-32, manganese-54, and radium-226 exhibit small seasonal variations (of the order of 0.1%), proposed to be related to either solar flare activity or distance from the sun. However, such measurements are highly susceptible to systematic errors, and a subsequent paper has found no evidence for such correlations in a half-dozen isotopes, and sets upper limits on the size of any such effects. However, research at Purdue University indicates that the rate of radioactive decay may not be truly constant, but slightly influenced by solar flares due to variations in solar neutrino flux.

Alpha Decay

Alpha decay is a type of radioactive decay in which an atomic nucleus emits an alpha particle and thereby transforms (or 'decays') into an atom with amass number 4 less and atomic number 2 less. Alpha decay is by far the most common form of cluster decay where the parent atom ejects a defined daughter collection of nucleons, leaving another defined product behind (in nuclear fission, a number of different pairs of daughters of approximately equal size are formed). Alpha decay is the most likely cluster decay because of the combined extremely high binding energy and relatively small mass of the helium-4 product nucleus (the alpha particle).

Alpha decay, like other cluster decays, is fundamentally a quantum tunneling process. Unlike beta decay, alpha decay is governed by the interplay between the nuclear force and the electromagnetic force. Alpha decay typically occurs in the heaviest nuclides. In theory it can occur only in nuclei somewhat heavier than nickel (element 28), where overall binding energy per nucleon is no longer a minimum, and the nuclides are therefore unstable toward spontaneous fission-type processes. In practice, this mode of decay has only been observed in nuclides considerably heavier than nickel, with the lightest known alpha emitter being the lightest isotopes (mass numbers 106–110) of tellurium (element 52).

Alpha particles have a typical kinetic energy of 5 MeV (that is, H" 0.13% of their total energy, i.e. 110 TJ/kg) and a speed of 15,000 km/s. This corresponds to a speed of around 0.05 *c*. There is surprisingly small variation around this energy, due to the heavy dependence of the half-life of this process on the energy produced. Because of their relatively large mass, +2 electric charge and relatively low velocity, alpha particles are very likely to interact with other atoms and lose their energy, so their forward motion is effectively stopped within a few centimeters of air. Most of the helium produced on Earth (approximately 99% of it) is the result of the alpha decay of underground deposits of minerals containing uranium or thorium. The helium is brought to the surface as a byproduct of natural gas production.

History

Alpha particles were first described in the investigations of radioactivity by Ernest Rutherford in 1899, and by 1907 they were identified as He^{2+} ions. For more details of this early work, see Alpha particle#History of discovery and use. By 1928, George Gamow had solved the theory of the alpha decay via tunneling. The alpha particle is trapped in a potential well by the nucleus. Classically, it is forbidden to escape, but according to the then newly discovered principles of quantum mechanics, it has a tiny (but non-zero) probability of "tunneling" through the barrier and appearing on the other side to escape the nucleus. Gamow solved a model potential for the nucleus and derived from first principles a relationship between the half-life of the decay, and the energy of the emission, which had been previously discovered empirically, and was known as the Geiger–Nuttall law.

Uses

Americium-241, an alpha emitter, is used in smoke detectors. The alpha particles ionize air between a small gap. A small current is passed through that ionized ar. Smoke particles from fire that enter the air gap reduce the current flow, sounding the alarm. Alpha decay can provide a safe power source for radioisotope thermoelectric generators used for space probes and artificial heart pacemakers. Alpha decay is much more easily shielded against than other forms of radioactive decay. Plutonium-238, for example, requires only 2.5 mm of lead shielding to protect against unwanted radiation. Static eliminators typically use polonium-210, an alpha emitter, to ionize air, allowing the 'static cling' to more rapidly dissipate.

Toxicity

Being relatively heavy and positively charged, alpha particles tend to have a very short mean free path, and quickly lose kinetic energy within a short distance of their source. This results in several MeV being deposited in a relatively small volume of material. This increases the chance of cellular damage in cases of internal contamination. In general, external alpha radiation is not harmful since alpha particles are effectively shielded by a few centimeters of air, a piece of paper, or the thin layer of dead skin cells.

Even touching an alpha source is usually not harmful, though many alpha sources also are accompanied by beta-emitting radio daughters, and alpha emission is also accompanied by gamma photon emission. If substances emitting alpha particles are ingested, inhaled, injected or introduced through the skin, then it could result in a measurable dose. The relative biological effectiveness (RBE) of alpha radiation is higher than that of beta or gamma radiation. RBE quantifies the ability of radiation to cause certain biological effects, notably either cancer or cell-death, for equivalent radiation exposure. The higher value for alpha radiation is generally attributable to the high linear energy transfer (LET), which is about one ionization of a chemical bond for every Ångström of travel by the alpha particle. The RBE has been set at the value of 20 for alpha radiation by various government regulations. The RBE is set at 10 for neutron irradiation, and at 1 for beta radiation and ionizing photons.

However, another component of alpha radiation is the recoil of the parent nucleus, termed alpha recoil. Due to the conservation of momentum requiring the parent nucleus to recoil, the effect acts much like the 'kick' of a rifle butt when a bullet goes in the opposite direction. This gives a significant amount of energy to the recoiling nucleus, which also causes ionization damage. The total energy of the recoil nucleus is readily calculable, and is roughly the weight of the alpha (4 u) divided by the weight of the parent (typically about 200 u) times the total energy of the alpha. By some estimates, this might account for most of the internal radiation damage, as the recoil nucli are typically heavy metals which preferentially collect on the chromosomes. In some studies, this has resulted in a RBE approaching 1,000 instead of the value used in governmental regulations.

The largest natural contributor to public radiation dose is radon, a naturally occurring, radioactive gas found in soil and rock. If the gas is inhaled, some of the radon particles may attach to the inner lining of the lung. These particles continue to decay, emitting alpha particles which can damage cells in the lung tissue. The death of Marie Curie at age 66 from leukemia was probably caused by prolonged exposure to high doses of ionizing radiation, but it is not clear if this was due to alpha radiation or X-rays. Curie worked extensively with radium, which decays into radon, along with other radioactive materials that emit beta and gamma rays. However, Curie also worked with unshielded X-ray tubes during World War I, and analysis of her skeleton during a reburial showed a relatively low level of radioisotope burden. Russian dissident Alexander Litvinenko's 2006 murder by radiation poisoning is thought to have been carried out with polonium-210, an alpha emitter.

Beta Decay

In nuclear physics, beta decy is a type of radioactive decay in which a beta particle (an electron or a positron) is emitted. In the case of electron emission, it is referred to as *beta minus* (β^-), while in the case of a positron emission as *beta plus* (β^+). In electron emission, an electron antineutrino is also emitted, while positron emission is accompanied by an electron neutrino. Beta decay is mediated by the weak force. Emitted beta particles have a continuous kinetic energy spectrum, ranging from 0 to the maximal available energy (Q), which depends on the parent and daughter nuclear states that participate in the decay. A typical Q is around 1 MeV, but it can range from a few keV to a few tens of MeV. Since the equivalence of energy of the rest mass of electron is 511 keV, the most energetic beta particles are ultrarelativistic, with speeds very close to the speed of light.

Sometimes electron capture decay is included as a type of beta decay (and is referred to as "inverse beta decay"), because the basic process, mediated by the weak force is the same. However, no beta particle is emitted, but only an electron neutrino. Instead of beta-plus emission, an inner atomic electron is captured by a proton in the nucleus. This type of decay is therefore analogus to positron emission (and also happens, as an alternative decay route, in all positron-emitters). However, the route of electron capture is the *only* type of decay that is allowed in proton-rich nuclides that *do not* have sufficient energy to emit a positron (and neutrino).

Double Beta Decay

Some nuclei can undergo double beta decay ($\beta\beta$ decay) where the charge of the nucleus changes by two units. Double beta decay is difficult to study in most practically interesting cases, because both β decay and $\beta\beta$ decay are possible, with probability favouring β decay; the rarer $\beta\beta$ decay process is masked by these events. Thus, $\beta\beta$ decay is usually studied only for beta stable nuclei. Like single beta decay, double beta decay does not change A; thus, at least one of the nuclides with some given A has to be stable with regard to both single and double beta decay. Beta decay can be considered as a perturbation as described in quantum mechanics, and thus follows Fermi's Golden Rule.

Bound-State β^- Decay

For fully ionized atoms (bare nuclei), it is possible for electrons to be emitted from the nucleus into low-lying atomic bound states (orbitals). This can not occur for neutral atoms whose low-lying bound states are already filled. The phenomenon was first observed for $^{163}Dy^{66+}$ in 1992 by Jung et al. of the Darmstadt Heavy-Ion Research group.

Although neutral ^{163}Dy is a stable isotope, the fully ionized $^{163}Dy^{66+}$ undergoes β decay into the K and L shells with a half-life of 47 days. Another possibility is that a fully ionized atom undergoes greatly accelerated β decay, as observed for ^{187}Re by Bosch et al., also at Darmstadt. Neutral ^{187}Re does undergo β decay with a half life of 42 x 10^9 years, but for fully ionized $^{187}Re^{75+}$ this is shortened by a factor of 10^9 to only 32.9 years. For comparison the variation of decay rates of other nuclear processes due to chemical environment is less than 1%.

Kurie Plot

A Kurie plot (also known as a Fermi-Kurie plot) is a graph used in studying beta decay developed by Franz N. D. Kurie, in which the square root of the number of beta particles whose momenta (or energy) lie within a certain narrow range, divided by a function worked out by Fermi, is plotted against beta-particle energy; it is a straight line for allowed transitions and some forbidden transitions, in accord with the Fermi beta-decay theory. Linear regression of a Fermi-Kurie Plot can help determining the maximum energy imparted to the electron/ positron by determining the energy-axis(x-axis) intercept.This graph helps us to know a better way of understanding of emission of beta particle.

Discovery and Characterization of b⁻ Decay

Radioactivity was discovered in 1896 by Henri Becquerel in uranium, and subsequently observed by Marie and Pierre Curie in thorium and in the new elements polonium and radium. In 1899 Ernest Rutherford separated radioactive emissions into two types: alpha and beta (now beta minus), based on penetration of objects and ability to cause ionization.

Alpha rays could be stopped by thin sheets of paper or aluminum, whereas beta rays could penetrate several millimetres of aluminum. (In 1900 Paul Villard identified gamma rays as a third, still more penetrating type of radiation.) In 1900 Becquerel measured the ratio of electric charge to mass (e/m) for beta particles by the method of J.J. Thomson used to study cathode rays and identify the electron. He found that e/m for a beta particle is the same as for Thomson's electron, and therefore suggested that the beta particle is in fact an electron.

In 1901 Rutherford and Frederick Soddy showed that alpha and beta radioactivity involves the transmutation of atoms into atoms of other chemical elements. In 1913, after the products of more radioactive decays were known, Soddy and Kazimierz Fajans independently proposed their radioactive displacement law, which states that beta (i.e. β^-) emission from one element produces another element one place to the right in the periodic table, while alpha emission produces an element two places to the left.

Neutrinos in Beta Decay

Historically, the study of beta decay provided the first physical evidence of the neutrino. In 1911 Lise Meitner and Otto Hahn performed an experiment that showed that the energies of electrons emitted by beta decay had a continuous rather than discrete spectrum. This was in apparent contradiction to the law of conservation of energy, as it appeared that energy was lost in the beta decay process. A second problem was that the spin of the Nitrogen-14 atom was 1, in contradiction to the Rutherford prediction of ½. In 1920-1927, Charles Drummond Ellis (along with James Chadwick and colleagues) established clearly that the beta decay spectrum is really continuous, ending all controversies. In a famous letter written in 1930 Wolfgang Pauli suggested that in addition to electrons and protons atoms also contained an extremely light neutral particle which he called the neutron. He suggested that this "neutron" was also emitted during beta decay and had simply not yet been observed. In 1931 Enrico Fermi renamed Pauli's "neutron" to neutrino, and in 1934 Fermi published a very successful model of beta decay in which neutrinos were produced.

Other Types of Beta Decay

In 1934 Frédéric and Irène Joliot-Curie bombarded aluminum with alpha particles to effect the nuclear reaction $^{4}He + ^{27}Al \longrightarrow ^{30}P + n$, and observed that the product isotope ^{30}P emits a positron identical to those found in cosmic rays by Carl David Anderson in 1932. This was the first example of β^- decay, which they termed artificial radioactivity since ^{30}P is a short-lived nuclide which does not exist in nature. The theory of electron capture was first discussed by Gian-Carlo Wick in a 1934 paper, and then developed by Hideki Yukawa and others. K-electron capture was first observed in 1937 by Luis Alvarez, in the nuclide ^{48}V. Alvarez went on to study electron capture in ^{67}Ga and other nuclides.

Gamma Ray

Gamma radiation, also known as gamma rays (denoted as γ), is electromagnetic radiation of high frequency (very short wavelength). They are produced by sub-atomic particle interactions such as electron-positron annihilation, neutral pion decay, fusion, fission or inverse Compton scattering in astrophysical processes. A classical gamma ray source is a type of radioactive decay called gamma decay where an excited nucleus emits a gamma ray almost immediately on formation. However, gamma decay may also describe isomeric transition which involves an inhibited gamma decay with a relatively much longer half life. Gamma rays have frequencies above 10 exahertz (10^{19} Hz), and therefore have energies above 100 keV and wavelength less than 10 picometers, less than the diameter of an atom.

Gamma rays from radioactive decay commonly have energies of a few hundred keV, and almost always less than 10 MeV. Energies from astronomical sources can be much higher, ranging over 10 TeV (this is far too large to result from radioactive decay). There is effectively no lower limit to gamma energy (though they are sometimes classed as X-rays if their frequencies are lower than 10^{19} Hz). Because gamma rays are a form of ionizing radiation, they pose a health hazard. Paul Villard, a French chemist and physicist, discovered gamma radiation in 1900, while studying radiation emitted from radium. Alpha and beta "rays" had already been separated and named by the work of Ernest Rutherford in 1899, and in 1903 Rutherford named Villard's distinct new radiation "gamma rays."

The distinction between X-rays and gamma rays has changed in recent decades. Originally, the electromagnetic radiation emitted by X-ray tubes had a longer wavelength than the radiation emitted by radioactive nuclei (gamma rays). Older literature distinguished between X- and gamma radiation on the basis of wavelength, with radiation shorter than some arbitrary wavelength, such as 10^{-11} m, defined as gamma rays. However, as shorter wavelength continuous spectrum "X-ray" sources such as linear accelerators and longer wavelength "gamma ray" emitters were discovered, the wavelength bands largely overlapped. The two types of radiation are now usually distinguished by their origin: X-rays are emitted by electrons outside the nucleus, while gamma rays are emitted by the nucleus.

Naming Conventions and Overlap in Terminology

In the past, the distinction between X-rays and gamma rays was based on energy (or equivalently frequency or wavelength), with gamma rays being considered a higher-energy version of X-rays. However, modern high-energy (megavoltage) X-rays produced by linear accelerators ("linacs") for megavoltage treatment, in cancer radiotherapy usually have higher energy (typically 4 to 25 MeV) than classical gamma rays produced by radioactive gamma decay.

Conversely, one of the most common gamma-ray emitting isotopes used in diagnostic nuclear medicine, technetium-99m, produces gamma radiation of about the same energy (140 keV) as produced by a diagnostic X-ray machine, and significantly lower energy than therapeutic photons from linacs.

Because of this broad overlap in energy ranges, the two types of electromagnetic radiation are now usually defined by their origin: X-rays are emitted by electrons (either in orbitals outside of the nucleus, or while being accelerated to produce Bremsstrahlung-type radiation), while gamma rays are emitted by the nucleus or from other particle decays or annihilation events.

There is no lower limit to the energy of photons produced by nuclear reactions, and thus ultraviolet and even lower energy photons produced by these processes would also be defined as "gamma rays". In certain fields such as astronomy, higher energy gamma and X-rays are still sometimes defined by energy, since the processes which produce them may be uncertain. Occasionally, high energy photons in nature which are known not to be produced by nuclear decay, are nevertheless referred to as gamma radiation. An example is "gamma rays" from lightning discharges at 10 to 20 MeV, which are known to be produced as Bremsstrahlung.

Units of Measure and Exposure

The measure of gamma rays' ionizing ability is called the exposure:

- The coulomb per kilogram (C/kg) is the SI unit of ionizing radiation exposure, and is the amount of radiation required to create 1 coulomb of charge of each polarity in 1 kilogram of matter.
- The röntgen (R) is an obsolete traditional unit of exposure, which represented the amount of radiation required to create 1 esu of charge of each polarity in 1 cubic centimeter of dry air. 1 röntgen $= 2.58\times10^{-4}$ C/kg

However, the effect of gamma and other ionizing radiation on living tissue is more closely related to the amount of energy deposited rather than the charge. This is called the absorbed dose:

- The gray (Gy), which has units of (J/kg), is the SI unit of absorbed dose, and is the amount of radiation required to deposit 1 joule of energy in 1 kilogram of any kind of matter.
- The rad is the (obsolete) corresponding traditional unit, equal to 0.01 J deposited per kg. 100 rad = 1 Gy.

The equivalent dose is the measure of the biological effect of radiation on human tissue. For gamma rays it is equal to the absorbed dose.

- The sievert (Sv) is the SI unit of equivalent dose, which for gamma rays is numerically equal to the gray (Gy).
- The rem is the traditional unit of equivalent dose. For gamma rays it is equal to the rad or 0.01 J of energy deposited per kg. 1 Sv = 100 rem.

Shielding

Shielding from gamma rays requires large amounts of mass. They are better absorbed by materials with high atomic numbers and high density, although neither effect is important compared to the total mass per area in the path of the gamma ray. For this reason, a lead shield is only modestly better (20–30%) as a gamma shield than an equal mass of another shielding material such as aluminium, concrete, or soil; lead's major advantage is its density.

The higher the energy of the gamma rays, the thicker the shielding required. Materials for shielding gamma rays are typically measured by the thickness required to reduce the intensity of the gamma rays by one half (the half value layer or HVL). For example gamma rays that require 1 cm (0.4") of lead to reduce their intensity by 50% will also have their intensity reduced in half by 4.1 cm of granite rock, 6 cm (2½") of concrete, or 9 cm (3½") of packed soil. However, the mass of this much concrete or soil is only 20–30% larger than that of lead with the same absorption capability. Depleted uranium is used for shielding in portable gamma ray sources, but again the savings in weight over lead is modest, and the main effect is to reduce shielding bulk.

Matter Interaction

When a gamma ray passes through matter, the probability for absorption in a thin layer is proportional to the thickness of that layer. This leads to an exponential decrease of intensity with thickness. The exponential absorption holds only for a narrow beam of gamma rays.

In passing through matter, gamma radiation ionizes via three main processes: the photoelectric effect, Compton scattering, and pai production.

- Photoelectric effect: This describes the case in which a gamma photon interacts with and transfers its energy to an atomic electron, ejecting that electron from the atom. The kinetic energy of the resulting photoelectron is equal to the energy of the incident gamma photon minus the binding energy of the electron. The photoelectric effect is the dominant energy transfer mechanism for X-ray and gamma ray photons with energies below 50 keV (thousand electron volts), but it is much less important at higher energies.
- Compton scattering: This is an interaction in which an incident gamma photon loses enough energy to an atomic electron to cause its ejection, with the remainder of the original photon's energy being emitted as a new, lower energy gamma photon with an emission direction different from that of the incident gamma photon. The probability of Compton scatter decreases with increasing photon energy. Compton scattering is thought to be the principal absorption mechanism for gamma rays in the intermediate energy range 100 keV to 10 MeV. Compton scattering is relatively independent of the atomic number of the absorbing material, which is why very dense metals like lead are only modestly better shields, on a *per weight* basis, than are less dense materials.
- Pair production: This becomes possible with gamma energies exceeding 1.02 MeV, and becomes important as an absorption mechanism at energies over about 5 MeV. By interaction with the electric field of a nucleus, the energy of the incident photon is converted into the mass of an electron-positron pair. Any gamma energy in excess of the equivalent rest mass of the two particles (1.02 MeV) appeas as the kinetic energy of the pair and the recoil nucleus. At the end of the positron's range, it combines with a free electron. The entire mass of these two particles is then converted into two gamma photons of at least 0.51 MeV energy each (or higher according to the kinetic energy of the annihilated particles).

The secondary electrons (and/or positrons) produced in any of these three processes frequently have enough energy to produce much ionization themselves.

Light Interaction

High-energy (from 80 to 500 GeV) gamma rays arriving from far far-distant quasars are used to estimate the extragalactic background light in the universe: The highest-energy rays interact more readily with the background light photons and thus their density may be estimated by analyzing the incoming gamma-ray spectrums.

Gamma Ray Production

Gamma rays from radioactive gamma decay are produced alongside other forms of radiation such as alpha or beta, and are produced after the other types of decay occur. The mechanism is that when a nucleus emits an α or β particle, the daughter nucleus is usually left in an excited state. It can then move to a lower energy state by emitting a gamma ray, in much the same way that an atomic electron can jump to a lower energy state by emitting infrared, visible, or ultraviolet light. Emission of a gamma ray from an excited nuclear state typically requires only 10^{-12} seconds, and is thus nearly instantaneous, following types of radioactive decay that produce other radioactive particles. Gamma decay from excited states may also happen rapidly following nuclear reactions such as nuclear fission or nuclear fusion.

In certain cases, the excited nuclear state following the emission of a beta particle may be more stable than average, and is termed a metastable excited state, if its decay is 100 to 1000 times longer than the average 10^{-12} seconds. Such nuclei have half lives that are easily measurable, and are termed nuclear isomers. Some (metastable) nuclear isomers are able to stay in their excited state for minutes, hours, days, or occasionally far longer, before emitting a gamma ry. Isomeric transition is the name given to a *gamma deay* from such a state. The proces of isomeric transition is therefore similar to any gamma emission, but differs in that it involves metastable excited states of the nuclei.

An emitted gamma ray from any type of excited state may transfer its energy directly to one of the most tightly bound electrons causing it to be ejected from the atom, a process termed the photoelectric effect (it should not be confused with the internal conversion process, in which no real gamma ray photon is produced as an intermediate particle). Gamma rays, X-rays, visible light, and radio waves are all forms of electromagnetic radiation. The only difference is the frequency and hence the energy of the photons. Gamma rays are generally the most energetic of these, although broad overlap with X-ray energies occurs. n optical spectroscopy, it is well known that an entity which emits light can also absorb light at the same wavelength (photon energy). For instance, a sodium flame can emit yellow light as well as absorb the yellow ight from a sodium vapor lamp. In the case of gamma rays, this can be seen in Mössbauer spectroscopy. Here, a correction for the energy lost by the recoil of the nucleus is made and the exact conditions for gamma ray absorption through resonance can be attained. This is similar to the Franck Condon effects seen in optical spectroscopy.

Health Effects

All ionizing radiation causes similar damage at a cellular level, but because rays of alpha particles and beta particles are relatively non-penetrating, external exposure to them causes only localized damage, e.g. radiation burns to the skin.

Gamma rays and neutrons are more penetrating, causing diffuse damage throughout the body (e.g. radiation sickness), increasing incidenc of cancer rather than burns. External radiation exposure should also be distinguished from internal exposure, due to ingested or inhaled radioactive substances, which, depending on the substance's chemical nature, can produce both diffuse and localized internal damage. The most biological damaging forms of gamma radiation occur in the gamma ray window, between 3 and 10 MeV, with higher energy gamma rays being less harmful because the body is relatively transparent to them. See cobalt-60.

Uses

Gamma rays travel to Earth across vast distances of the universe, only to be absorbed by Earth's atmosphere. Dfferent wavelengths of light penetrate Earth's atmosphere to different depths. Instruments aboard high-altitude balloons and such satellites as the Compton Observatory provide our only view of the gamma-spectrum sky. Gamma-induced molecular changes can also be used to alter the properties of semi-precious stones, and is often used to change white topaz into blue topaz.

In the US, gamma ray detectors are beginning to be used as part of the Container Security Initiative (CSI). These US$5 million machines are advertised to scan 30 containers per hour. The objective of this technique is to screen merchant ship containers before they enter US ports. Gamma radiation is often used to kill living organisms, in a process called irradiation. Applications of this include sterilizing medical equipment (as an alternative to autoclaves or chemical means), removing decay-causing bacteria from many foods or preventing fruit and vegetables from sprouting to maintain freshness and flavor. Despite their cancer-causing properties, gamma rays are also used to treat some types of cancer, since the rays kill cancer cells also. In the procedure called gamma-knife surgery, multiple concentrated beams of gamma rays are directed on the growth in order to kill the cancerous cells. The beams are aimed from different angles to concentrate the radiation on the growth while minimizig damage to surrounding tissues. Gamma rays are also used for diagnostic purposes in nuclear medicine in imaging techniques. A number of different gamma-emitting radioisotopes are used.

For example, in a PET scan a radiolabled sugar called fludeoxyglucose emits positrons that are converted to pairs of gamma rays that localize cancer (which often takes up more sugar than other surrounding tissues). The most common gamma emitter used in medical applications is the nuclear isomer technetium-99m which emits gamma rays in the same energy range as diagnostic X-rays. When this radionuclide tracer is administered to a patient, a gamma camera can be used to form an image of the radioisotope's distribution by detecting the gamma radiation emitted. Depending on what molecule has been labeled with the tracer, such techniques can be employed to diagnose a wide range of conditions (for example, the spread of cancer to the bones in a bone scan).

Body Response

When gamma radiation breaks DNA molecules, a cell may be able to repair the damaged genetic material, within limits. However, a study of Rothkamm and Lobrich has shown that this repair process works well after high-dose exposure but is much slower in the case of a low-dose exposure.

Risk Assessment

The natural outdoor exposure in Great Britain ranges from 2×10^{-7} to 4×10^{-7} cSv/h (centisieverts per hour). Natural exposure to gama rays is about 0.1 to 0.2 cSv per year, and the average total amount of radiation received in one year per inhabitant in the USA is 0.36 cSv. There is a small increase in the dose, due to naturally occurring gamma-radiation, around small particles of high atomic number materials in the human body caused by the photoelectric effect. By comparison, the radiation dose from chest radiography is a fraction of the annual naturally occurring background radiation dose, and the dose from fluoroscopy of the stomach is, at most, 5 cSv on the skin of the back. For acute full-body equivalent dose, 100 cSv causes slight blood changes, but 200–350 cSv causes very severe syndrome of nausea, hair loss, and hemorrhaging, and will cause death in a sizable number of cases—about 10% to 35% without medical treatment. A dose of 500 cSv is considered approximately the LD_{50} (lethal dose for 50% of exposed population) for an acute exposure to radiation even with standard medical treatment. A dose higher than 500 cSv brings an increasing chance of death above 50%.

Above 750–1000 cSv, even extraordinary treatment, such as bone-marrow transplants, will not prevent the death of the individual exposed. For low dose exposure, for example among nuclear workers, who receive an average yearly radiation dose of 1.9 cSv, the risk of dying from cancer (excluding leukemia) increases by 2 percent. For a dose of 10 cSv, that risk increase is at 10 percent. By comparison, risk of dying from cancer was increased by 32 percent for the survivors of the atomic bombing of Hiroshima and Nagasaki.

Double Beta Decay

Double beta decay is a radioactive decay process where a nucleus releases two beta rays as a single process. In double-beta decay, two neutrons in the nucleus are converted to protons, and two electrons and two electron antineutrinos are emitted. In the process ofbeta minus decay, unstable nuclei decay by converting a neutron in the nucleus to a proton and emitting an electron and an electron antineutrino. In order for beta decay to be possible, the final nucleus must have a larger binding energy than the original nucleus.

For some nuclei, such as germanium-76, the nucleus with atomic number one higher has a smaller binding energy, preventing beta decay from occurring. However, the nucleus with atomic number two higher, selenium-76, has a larger binding energy, so the double-beta decay process is allowed. Although efforts to observe the process date back to 1948, it was not observed in a laboratory setting until 1986. Geochemical observation of the decay products (by extraction of Kr and Xe from very old Se and Te minerals) are known since 1950. Double-beta decay is the rarest known kind of radioactive decay; it has been observed for only 11 isotopes, and all of them have a mean lifetime of more than 10^{19} yr.

For some nuclei, the process occurs as conversion of two protons to neutrons, with emission of two electron neutrinos and absorption of two orbital electrons (double electron capture). If the mass difference between the parent and daughter atoms is more than 1.022 MeV/c^2 (two electron masses), another decay branch is accessible, with capture of one orbital electron and emission of one positron. When the mass difference is more than 2.044 MeV/c^2 (four electron masses), the emission of two positrons is possible. These theoretically possible double-beta decay branches have not yet been observed.

Neutrinoless Double-Beta Decay

The processes described in the previous section are also known as two neutrino double-beta decay, as two neutrinos (or antineutrinos) are emitted. If the neutrino is a Majorana particle (meaning that the antineutrino and the neutrino are actually the same particle), and at least one type of neutrino has non-zero mass (which has been established by the neutrino oscillation experiments), then it is possible for neutrinoless double-beta decay to occur. In the simplest theoretical treatment of neutrinoless double-beta decay (light neutrino exchange), in essence the two neutrinos annihilate each other, or equivalently, one nucleon absorbs the neutrino emitted by another nucleon of the nucleus. The neutrinos in the above diagram are virtual particles. With only the two electrons in the final state, the total kinetic energy of the two electrons would be approximately the difference in binding energy between the initial and final state nuclei (with the recoil of the nucleus accounting for the rest of the available energy). To a very good approximation, the two electrons are emitted back-to-back to conserve momentum.

Neutrinoless Double-Beta Decay Experiments

Numerous experiments have been carried out to search for neutrinoless double-beta decay. Some recent and proposed future experiments include:

- Completed experiments:
  - Gotthard TPC
  - Heidelberg-Moscow
  - IGEX
  - NEMO1 and 2
- Current and proposed experiments
  - COBRA
  - CUORICINO and CUORE
  - EXO
  - GERDA
  - MAJORANA
  - MOON
  - NEMO-3 and SuperNEMO
  - SNO+

Klapdor-Kleingrothaus et al.'s Claimed Observation of Neutrinoless Double-Beta Decay

A subgroup of the Heidelberg-Moscow collaboration led by H. V. Klapdor-Kleingrothaus published a series of controversial papers claiming to have observed neutrinoless double-beta decay in 76

Ge with statistical significance of > 99.9999998%. This claim has been criticized by many in the nuclear and particle physics communities, but Klapdor-Kleingrothaus *et al.* have vigorously defended their work. Their measured value of $m_{\beta\beta}$ between ~0.2-0.5 eV (depending on the choice of nuclear matrix element) has yet to be confirmed by other experiments.

Double Electron Capture

Double electron capture is a decay mode of atomic nucleus. For a nuclide (*A*, *Z*) with number of nucleons *A* and atomic number *Z*, double electron capture is only possible if the mass of the nuclide of (*A*, *Z*-2) is lower. In this mode of decay, two of the orbital electrons are captured by two protons in the nucleus, forming two neutrons. Two neutrinos are emitted in the process.

Since the protons are changed to neutrons, the number of neutrons increases by 2, the number of protons *Z* decreases by 2, and the atomic mass number *A* remains unchanged. By changing the number of protons, double electron capture transforms the nuclide into a new element. In most cases this decay mode is masked by more probable modes (single electron capture etc.), but when all these modes are forbidden or strongly suppressed, double electron capture becomes the main mode of decay.

There exist 35 naturally occurring isotopes that can undergo double electron capture. However, there are no confirmed observations of this process. One reason is that the probability of double electron capture is enormously small (the theoretical predictions of half-lives for this mode lies well above 10^{20} years). A second reason is that the only detectable particles created in this process are X-rays and Auger electrons that are emitted by the excited atomic shell. In the range of their energies (~1-10 keV), the background is usually high. Thus, the experimental detection of double electron capture is more difficult than that for double beta decay. If the mass difference between te mother and daughter atoms is more than two masses of electron (1.022 MeV), the energy released in the process is enough to allow another mode of decay: electron capture with positron emission. It occurs simultaneously with double electron capture, their branching ratio depending on nuclear properties. When the mass difference is more than four electron masses (2.044 MeV), the third mode - double positron decay - is allowed. Only 6 naturally occurring nuclides can decay via these three modes simultaneously.

Neutrinoless Double Electron Capture

The above described process with capture of two electrons and emission of two neutrinos (two-neutrino double electron capture) is allowed by the Standard Model of particle physics: no conservation laws (including lepton number conservation) are violated. However, if the lepton number is not conserved, another kind of the process can occur: the so-called neutrinoless double electron capture. In this case, two electrons are captured by nucleus, but neutrinos are not emitted. The energy released in this process is carried away by an internal bremsstrahlung gamma quantum.

Internal Conversion

Internal conversion is a radioactive decay process where an excited nucleus interacts with an electron in one of the lower atomic orbitals, causing the electron to be emitted from the atom. Thus, in an internal conversion process, a high-energy electron is emitted from the radioactive atom, but without beta decay taking place. For this reason, the high-speed electrons from internal conversion are not beta particles (β particles), since the latter come from beta decay. Since no beta decay takes place in internal conversion, the element atomic number does not change, and thus (as is the case with gamma decay) no transmutation of one element to another is seen. Also, no neutrino is emitted in internal conversion. Internally converted electrons do not have the characteristic energetically-spread spectrum of β particles, which results from varying amounts of decay-energy being carried off by the neutrino (or antineutrino) in beta decay. Internally converted electrons, which carry a fixed fraction of the characteristic decay energy, have a well-specified discrete energy. The energy spectrum of a β particle is thus a broad hump, extending to a maximum decay energy value, while the spectrum of internally converted electrons is a sharp peak.

Mechanism

In the internal conversion process, the wavefunction of an inner shell electron penetrates the nucleus (i.e. there is a finite probability of the electron in an s atomic orbital being found in the nucleus) and when this is the case, the electron may couple to the excited state and take the energy of the nuclear transition directly, without an intermediate gamma ray being produced first. As an electromagnetic quantum process, the process of imparting energy to the electron may be seen as taking place by means of a virtual photon, but in that sense the photon involved can be considered as a "virtual gamma ray", which never appears except as a feature of an equation, rather than a directly measurable particle.

The kinetic energy of the emitted electron is equal to the transition energy in the nucleus, minus the binding energy of the electron. Most internal conversion electrons come from the K shell (1s state, see electron shell), as these two electrons have the highest probability of being found inside the nucleus. After the electron has been emitted, the atom is left with a vacancy in one of the inner electron shells. This hole will be filled with an electron from one of the higher shells and subsequently a characteristic x-ray or Auger electron will be emitted.

When the Process is Expected

Internal conversion is favoured when the energy gap between nuclear levels is small, and is also the primary mode of de-excitation for $0^+ \longrightarrow 0^+$ (i.e. E0) transitions (i.e., where excited nuclei are able to rid themselves of energy without changing electric and magnetic moments in certain ways) with insufficient energy to decay by pair production. It is the predominant mode of de-excitation whenever the initial and final spin states are the same, but the multi-polarity rules for nonzero initial and final spin states do not necessarily forbid the emission of a gamma ray in such a case. The tendency towards internal conversion can be determined by the internal conversion coefficient, which is empirically determined by the ratio of de-excitations that go by the emission of electrons to those that go by gamma emission. The internal conversion process competes with gamma decay. This competition is quantified in the form of the internal conversion coefficient which is defined as $\alpha = e / \gamma$ where e is the rate of conversion electrons and γ is the rate of gamma-ray emission observed from a decaying nucleus. For example, in the decay of an excited state of the nucleus of ^{125}I, 7% of the decays emit energy as a gamma ray, while 93% release energy as conversion electrons.

Therefore, this excited state of ^{125}I has an internal conversion coefficient of $\alpha = 13.6$. Internal conversion coefficients are observed to increase for increasing atomic number (Z) and decreasing gamma-ray energy. As one example, IC coefficients are calculated explicitly for ^{55}Fe, ^{67}Ga, ^{99m}Tc, ^{111}In, ^{113m}In, ^{115m}In, ^{123}I, ^{125}I, ^{193m}Pt, ^{201}Tl and ^{203}Pb by Howell (1992) using Monte Carlo methods (for ^{55}Fe the IC coefficient is zero). The energy of the emitted gamma-ray is regarded as a precise measure of the difference in energy between the excited states of the decaying nucleus. However, this is not true in the case of conversion electrons. The energy of a conversion electron is given as $E = (E_i - E_f) - E_B$ where E_i and E_f are the energies of the nucleus in its initial and final states, respectively, while E_B is the binding energy of the electron.

Similar Processes

This internal conversion process is also not to be confused with the similar photoelectric effect, which also may occur with gamma radiation associated electron emission, in which an incident gamma photon emitted from a nucleus interacts with an electron, expelling the electron from the atom. Thus, gamma photoelectric effect electron emission may also cause high-speed electrons to be emitted from radioactive atoms without beta decay.

However, in internal conversion, the nucleus does not first emit an intermediate real gamma ray, and therefore need not change angular momentum or electric moment. Also, electrons from the gamma photoelectric effect show a spread in energy, depending on how much energy has been imparted to the ejected electron by the gamma ray which interacts with it—an amount which is variable depending on the angle of gamma photon scattering from the electron. Further, a gamma ray is still emitted in photoelectric processes, but one which possesses a fraction of the energy than the gamma ray which left the nucleus. By contrast, in internal conversion, as noted, no gamma ray is emitted at all and the electron energy is fixed at a single, typical value. Auger electrons, which may also be produced after an internal conversion, arise from a mechanism that is different from that of internal conversion, but is analogous to it. Internal conversion electrons arise when an intense electric dipole field inside the nucleus accelerates an electron which has penetrated the nucleus, to remove it from the atom. Auger electrons similarly arise when an electric field is produced within an atom's electron cloud due to loss of another electron, and this field again induces the acceleration and removal of yet another of the atom's atomic orbital electrons. Like internal conversion electrons, Auger electrons also emerge in a sharp energy peak. The electron capture (EC) process also involves an inner shell electron, which in this case is retained in the nucleus (changing the atomic number) and leaving the atom (not the nucleus) in an excited state. The atom can relax by X-ray emission and/or by Auger electron emission. Unstable nuclei can usually decay through both IC and EC processes.

Isomeric Transition

An isomeric transition is a radioactive decay process that involves emission of a gamma ray from an atom where the nucleus is in an excited metastable state, referred to in its excited state, as a nuclear isomer. The emission of a gamma ray from an excited nuclear state allows the nucleus to lose energy and reach either a lower energy state, sometimes its ground state. In certain cases, the excited nuclear state following a nuclear reaction or other type of radioactive decay, has a half live that is more than 100 to 1000 times longer than the average 10^{-12} seconds, and this excited state is referred to as a metastable nuclear excited state.

Some nuclei are able to stay in this metastable excited state for minutes, hours, days, or occasionally far longer, before undergoing *gamma decay*, in which they undergo radioactive decay with primary emission of a gamma ray. The process of isomeric transition (that is, the gamma decay of so-called nuclear isomers), is therefore similar to any gamma emission from any excited nuclear state, but differs in that it involves excited metastable states of nuclei with longer half lives. These states are created, as in all nuclei that undergo gamma radioactive decay, following the emission of an alpha, beta particle, or occasionally other types of particles that leave the nucleus in an excited state.

The gamma ray may transfer its energy directly to one of the most tightly bound electrons causing it to be ejected from the atom, a process termed the photoelectric effect. This should not be confused with the internal conversion process, in which no gamma ray photon is produced as an intermediate particle.

Cluster Decay

Cluster decay (also named heavy particle radioactivity or heavy ion radioactivity) is a type of nuclear decay in which a parent atomic nucleus with A nucleons and Z protons emits a cluster of N_e neutrons and Z_e protons heavier than an alpha particle but lighter than a typical binary fission fragment (although ternary fission into three fragments produces products which overlap cluster decay). Cluster decay, like alpha decay, is fundamentally a quantum tunneling process: in order to be emitted the cluster must penetrate a potential barrier. This is a different process than that which preceeds light fragment emission in ternary fission, which results from previous nuclear deformation and energy input, as a result of a nuclear reaction (most often from a neutron).

By contrast, cluster emission is a type of spontaneous radioactive decay. Theoretically any nucleus with Z > 40 for which the released energy (Q value) ia a positive quantity can be a cluster-emitter. In practice, observations are severely restricted to limitations imposed by currently available experimental techniques which require a sufficiently short half-life, $T_c < 10^{32}$ s, and a sufficiently large branching ratio $B > 10^{-17}$. In the absence of any energy loss for fragment deformation and excitation, as in cold fission phenomena or in alpha decay, the total kinetic energy is equal to the Q-value and is divided between the particles in inverse proportion with their masses, as required by conservation of linear momentum

$$E_k = QA_d / A$$

where A_d is the mass number of the daughter, $A_d = A - A_e$.

Cluster decay exists in an intermediate position between alpha decay (in which a nucleus spits out a He^4 nucleus), and spontaneous fission, in which a heavy nucleus splits into two (or more) large fragments and an assorted number of neutrons. Spontaneous fission ends up with a probabilistic distribution of daughter products, which sets it apart from cluster decay. In cluster decay for a given radioisotope, the emitted particle is a light nucleus and the decay method always emits this same particle. Nevertheless, for heavier emitted clusters there is practically no qualitative difference between cluster decay and spontaneous cold fission.

History

The first informations about the atomic nucleus were obtained at the beginning of the 20th century by studying the radioactivity. For a long period of time only three kinds of nuclear decay modes (alpha, beta and gamma) have been known. They illustrate three of the fundamental interactions in nature: strong; weak, and electromagnetic.

Spontaneous fission became popular soon after its discovery in 1940 by K. Petrzhak and G. Flerov owing to both military and peaceful applications of the neutron-induced fission discovered in 1939 by O. Hahn, Lise Meitner and F. Strassmann, employing the large amount of energy released during the process. At present we know many other kinds of radioactivities, e.g. cluster decay, proton decay, various beta-delayed decay modes (p, 2p, 3p, n, 2n, 3n, 4n, d, t, alpha, f), fission isomers, particle accompanied (ternary) fission, etc. The height of the potential barrier, mainly of Coulomb nature, for emission of the charged particles is much higher than the observed kinetic energy of the emitted particles. The spontaneous decay can only be explained by quantum mechanical tunneling in a similar way to the first application of the Quantum Mechanics to Nuclei given by G. Gamow for alpha decay.

> *"In 1980 A. Sandulescu, D.N. Poenaru, and W. Greiner described calculations indicating the possibility of a new type of decay of heavy nuclei intermediate between alpha decay and spontaneous fission. The first observation of heavy-ion radioactivity was that of a 30-MeV, carbon-14 emission from radium-223 by H.J. Rose and G.A. Jones in 1984".*

Usually the theory explains an already experimentally observed phenomenon. Cluster decay is one of the rare examples of phenomena predicted before experimental discovery. Theoretical predictions were made in 1980, four years before experimental discovery. Four theoretical approaches were used: fragmentation theory by solving a Schroedinger equation with mass assymetry as a variable to obtain the mass distributions of fragments; penetrability calculations similar to those used in traditional theory of alpha decay, and superasymmetric fission models, numerical (NuSAF) and analytical (ASAF). Superasymmetric fission models are based on the macroscopic-microscopic approach using the asymmetrical two-center shell model level energies as input data for the shell and pairing corrections. Either the liquid drop model or the Yukawa-plus-exponential model extended to different charge-to-mass ratios have been used to calculate the macroscopic deformation energy.

Experiments

The main experimental difficulty in observing cluster decay comes from the need to identify a few rare events among an enormous number of background alpha particle. The quantities experimentally determined are the partial halflife, T_c, and the kinetic energy of the emitted cluster E_k. There is also a need to identify the emitted particle. Detection of radiations is based on their interactions with matter, leading mainly to ionizations.

Using a semiconductor telescope and conventional electronics to identify the ^{14}C ions, the Rose and Jones's experiment was running for about six months in order to get 11 useful events. With modern magnetic spectrometers (SOLENO and Enge-split pole), at Orsay and Argonne National Laboratory, a very strong source could be used, so that results were obtained in a run of few hours. Solid state nuclear track detectors (SSNTD) insensitive to alpha particles and magnetic spectrometers in which alpha particles are deflected by a strong magnetic field have been used to overcome this difficulty. SSNTD are cheap and handy but they need chemical etching and microscope scanning.

Fine Structure

The fine structure in 14-C radioactivity of 223-Ra was discussed for the first time by M. Greiner and W. Scheid in 1986. The superconducting spectrometer SOLENO of IPN Orsay has been used since 1984 to identify ^{14}C clusters emitted from $^{222\text{-}224,226}Ra$ nuclei. Moreover it was used to discover the fine structure observing transitions to excited states of the daughter. A transition with an excited state of ^{14}C predicted in Ref. was not yet observed. Surprisingly, the experimentalists had seen a transition to the first excited state of the daughter stronger than that to the ground state. The transition is favoured if the uncoupled nucleon is left in the same state in both parent and daughter nuclei. Otherwise the difference in nuclear structure leads to a large hindrance. The interpretation was confirmed: the main spherical component of the deformed parent wave function has an i11/2 character, i.e. the main component is spherical.

Chapter 5

Nuclear Energy

Dissipation

Dissipation embodies the concept of a dynamical system where important mechanical models, such as waves or oscillations, lose energy over time, typically from friction or turbulence. The lost energy converts into heat, which raises the temperature of the system. Such systems are called dissipative systems. For example, a wave that lses amplitude is said to dissipate.

The precise nature of the effects depends on the nature of the wave: an atmospheric wave, for instance, may dissipate close to the surface due to friction with the land mass, and at higher levels due to radiative cooling. Dissipating forces are those that can not be described by Hamiltonian formalism. Loosely speaking, this includes friction, and all similar forces that result in decoherency of energy—that is, conversion of coherent or directed energy flow into an indirected or more isotropic distribution of energy. In computational physics, numerical dissipation (also known s "numerical diffusion") refers to certain side-effects that may occur as a result of a numerical solution to a differential equation.

When the pure advection equation, which is free of dissipation, is solved by a numerical approximation method, the energy of the initial wave may be reduced in a way analogous to a diffusional process. Such a method is said to contain 'dissipation'. In some cases, "artificial dissipation" is intentionally added to improve the numerical stability characteristics of the solution. A formal, mathematical definition of dissipation, as commonly used in the mathematical study of measure-preserving dynamical systems, is given in the article *wandering set*.

In River Hydrology

Dissipation is the process of converting mechanical energy of downward-flowing water into thermal and acoustical energy. Various devices are designed in streambeds to reduce the kinetic energy of flowing waters to reduce their erosive potential on banks and river bottoms. Very often these devices look like small waterfalls or cascades, where water flows vertically or over riprap to lose some of its kinetic force.

Photovoltaic Effect

The photovoltaic effect is the creation of a voltage or a corresponding electric current in a material upon exposure to light. Though the photovoltaic effect is directly related to the photoelectric effect, the two processes are different and should be distinguished. In the photoelectric effect, electrons are ejected from a material's surface upon exposure to radiation of sufficient energy. The photovoltaic effect is different in that the generated electrons are transferred between different bands (i.e., from the valence to conduction bands) within the material, resulting in the buildup of a voltage between two electrodes. In most photovoltaic applications the radiation is sunlight and for this reason the devices are known as solar cells. In the case of a p-n junction solar cell, illuminating the material creates an electric current as excited electrons and the remaining holes are swept in different directions by the built-in electric field of the depletion region. The photovoltaic effect was first observed by Alexandre-Edmond Becquerel in 1839.

Photoelectric Effect

In the photoelectric effect, electrons are emitted from matter (metals and non-metallic solids, liquids or gases) as a consequence of their absorption of energy from electromagnetic radiation of very short wavelength, such as visible or ultraviolet light. Electrons emitted in this manner may be referred to as "photoelectrons". First observed by Heinrich Hertz in 1887, the phenomenon is also known as the "Hertz effect", although the latter term has fallen out of general use. Hertz observed and then showed that electrodes illuminated with ultraviolet light create electric sparks more easily. The photoelectric effect requires photons with energies from a few electronvolts to over 1 MeV in high atomic number elements. Study of the photoelectric effect led to important steps in understanding the quantum nature of light and electrons and influenced the formation of the concept of wave–particle duality. Other phenomena where light affects the movement of electric charges include the photoconductive effect (also known as photoconductivity or photoresistivity), the photovoltaic effect, and the photoelectrochemical effect.

Emission Mechanism

The photons of a light beam have a characteristic energy determined by the frequency of the light. In the photoemission process, if an electron within some material absorbs the energy of one photon and thus has more energy than the wor function (th electron bindng energy) of the mateial, it is ejeted. If the photon energy is too low, the electron is unable to escape the material. Increasing the intensity of the light beam increases the number of photons in the light beam, and thus increases the number of electrons excited, but does not increase the energy that each electron possesses.

The energy of the emitted electrons does not depend on the intensity of the incoming light, but only on the energy or frequency of the individual photons. It is an interaction between the incident photon and the outermost electron. Electrons can absorb energy from photons when irradiated, but they usually follow an "all or nothing" principle. All of the energy from one photon must be absorbed and used to liberate one electron from atomic binding, or else the energy is re-emitted. If the photon energy is absorbed, some of the energy liberates the electron from the atom, and the rest contributes to the electron's kinetic energy as a free particle.

Experimental Results of the Photoelectric Emission

1. For a given metal and frequency of incident radiation, the rate at which photoelectrons are ejected is directly proportional to the intensity of the incident light.
2. For a given metal, there exists a certain minimum frequency of incident radiation below which no photoelectrons can be emitted. This frequency is called the threshold frequency.
3. For a given metal of particular work function, increase in intensity of incident beam increases the magnitude of the photoelectric current, though stoppage voltage remains the same.
4. For a given metal of particular work function, increase in frequency of incident beam increases the maximum kinetic energy with which the photoelectrons are emitted, but the photoelectric current remains the same, though stoppage voltage increases.

5. Above the threshold frequency, the maximum kinetic energy of the emitted photoelectron depends on the frequency of the incident light, but is independent of the intensity of the incident light so long as the latter is not too high
6. The time lag between the incidence of radiation and the emission of a photoelectronis very small, less than 10^{-9} second.
7. The direction of distribution of emitted electrons peaks in the direction of polarization (the direction of the electric field) of the incident light, if it is linearly polarized.

Stopping Potential

The relation between current through an illuminated photoelectric system and applied voltage illustrates the nature of the photoelectric effect. For discussion, a plate P is illuminated by a light source, and any emitted electroncs are collected at another plate electrode Q. The potential between P and Q can be varied and the current flowing in the external circuit between P and Q is measured. If the frequency and the intensity of the incident radiation are kept fixed, it is found that the photoelectric current increases gradually with the increase in positive potential until all the photoelectrons emitted are collected. The photoelectric current attains saturation value and it does not increase further for any increase in the positive potential.

The saturation current depends on the intensity of illumination, but not its wavelength. If we apply negative potential to plate Q with respect to plate P, and increases it gradually we note that photoelectric current decreases rapidly until it is zero, at a certain negative potential on plate Q.The minimum negative potential given to plate Q at which the photoelectric current becomes zero is called stopping potential or cut off potential. i. For the given frequency of incident radiation, the stopping potential is independent of its intensity. ii. For a given frequency of the incident radiation, the stopping potential V0 if related to the maximum kinetic energy of the photoelectron that is just stopped from reaching plate Q. The stopping voltage varies linearly with frequency of light, but depends on the type of material. For any particular material, there is a threshold frequency that must be exceeded, independant of light intensity, to observe any electron emission.

Three-Step Model

In the X-ray regime, the photoelectric effect in crystalline material is often decomposed into three steps:

1. Inner photoelectric effect. The hole left behind can give rise to auger effect, which is visible even when the electron does not leave the material. In molecular solids phonons are excited in this step and may be visible as lines in the final electron energy. The inner photoeffect has to be dipole allowed. The transition rules for atoms translate via the tight-binding model onto the crystal. They are similar in geometry to plasma oscillations in that they have to be transversal.
2. Ballistic transport of half of the electrons to the surface. Some electrons are scattered.
3. Electrons escape from the material at the surface.

In the three-step model, an electron can take multiple paths through these three steps. All paths can interfere in the sense of the path integral formulation. For surface states and molecules the three-step model does still make some sense as even most atoms have multiple electrons which can scatter the one electron leaving.

History

When a surface is exposed to electromagnetic radiation above a certain threshold frequency (typically visible light for alkali metals, near ultraviolet for other metals, and extreme ultraviolet for non-metals), the radiation is absorbed and electrons are emitted. Light, and especially ultra-violet light, discharges negatively electrified bodies with the production of rays of the same nature as cathode rays.

Under certain circumstances it can directly ionize gases. The first of these phenomena was discovered by Hertz and Hallwachs in 1887. The second was announced first by Lenard in 1900. The ultra-violet light to produce these effects may be obtained from an arc lamp, or by burning magnesium, or by sparking with an induction coil between zinc or cadmium terminals, the light from which is very rich in ultra-violet rays.

Sunlight is not rich in ultra-violet rays, as these have been absorbed by the atmosphere, and it does not produce nearly so large an effect as the arc-light. Many substances besides metals discharge negative electricity under the action of ultraviolet light: lists of these substances will be found in papers by G. C. Schmidt and O. Knoblauch.

19th Century

In 1839, Alexandre Edmond Becquerel discovered the photovoltaic effect while studying the effect of light on electrolytic cells. Though not equivalent to the photoelectric effect, his work on photovoltaics was instrumental in showing a strong relationship between light and electronic properties of materials. In 1873, Willoughby Smith discovered photoconductivity in selenium while testing the metal for its high resistance properties in conjunction with his work involving submarine telegraph cables. Johann Elster (1854–1920) and Hans Geitel (1855–1923), students in Heidelberg, developed the first practical photoelectric cells that could be used to measure the intensity of light. Elster and Geitel had investigated with great success the effects produced by light on electrified bodies.

In 1887, Heinrich Hertz observed the photoelectric effect and the production and reception of electromagnetic waves. He published these observations in the journal Annalen der Physik. His receiver consisted of a coil with a spark gap, where a spark would be seen upon detection of electromagnetic waves. He placed the apparatus in a darkened box to see the spark better.

However, he noticed that the maximum spark length was reduced when in the box. A glass panel placed between the source of electromagnetic waves and the receiver absorbed ultraviolet radiation that assisted the electrons in jumping across the gap. When removed, the spark length would increase. He observed no decrease in spark length when he substituted quartz for glass, as quartz does not absorb UV radiation. Hertz concluded his months of investigation and reported the results obtained. He did not further pursue investigation of this effect.

The discovery by Hertz in 1887 that the incidence of ultra-violet light on a spark gap facilitated the passage of the spark, led immediately to a series of investigations by Hallwachs, Hoor, Righi and Stoletow. on the effect of light, and especially of ultra-violet light, on charged bodies. It was proved by these investigations that a newly cleaned surface of zinc, if charged with negative electricity, rapidly loses this charge however small it may be when ultra-violet light falls upon the surface; while if the surface is uncharged to begin with, it acquires a positive charge when exposed to the light, the negative electrification going out into the gas by which the metal is surrounded; this positive electrification can be much increased by directing a strong airblast against the surface. If however the zinc surface is positively electrified it suffers no loss of charge when exposed to the light: this result has been questioned, but a very careful examination of the phenomenon by Elster and Geitel has shown that the loss observed under certain circumstances is due to the discharge by the light reflected from the zinc surface of negative electrification on neighbouring conductors induced by the positive charge, the negative electricity under the influence of the electric field moving up to the positively electrified surface. With regard to the *Hertz effect*, the researches from the start showed a great complexity of the phenomenon of photoelectric fatigue — that is, the progressive diminution of the effect observed upon fresh metallic surfaces. According to an important research by Wilhelm Hallwachs, ozone played an important part in the phenomenon. However, other elements enter such as oxidation, the humidity, the mode of polish of the surface, etc. It was at the time not even sure that the fatigue is absent in a vacuum.

In the period from February 1888 and until 1891, a detailed analysis of photoeffect was performed by Aleksandr Stoletov with results published in 6 works; four of them in *Comptes Rendus*, one review in *Physikalische Revue* (translated from Russian), and the last work in *Journal de Physique*. First, in these works Stoletov invented a new experimental setup which was more suitable for a quantitative analysis of photoeffect. Using this setup, he discovered the direct proportionality between the intensity of light and the induced photo electric current (the first law of photoeffect or Stoletov's law).

One of his other findings resulted from measurements of the dependence of the intensity of the electric photo current on the gas pressure, where he found the existence of an optimal gas pressure P_m corresponding to a maximum photocurrent; this property was used for a creation of solar cells. In 1899, J. J. Thomson investigated ultraviolet light in Crookes tubes. Influenced by the work of James Clerk Maxwell, Thomson deduced that cathode rays consisted of negatively charged particles, later called electrons, which he called "corpuscles". In the research, Thomson enclosed a metal plate (a cathode) in a vacuum tube, and exposed it to high frequency radiation. It was thought that the oscillating electromagnetic fields caused the atoms' field to resonate and, after reaching a certain amplitude, caused a subatomic "corpuscle" to be emitted, and current to be detected. The amount of this current varied with the intensity and colour of the radiation. Larger radiation intensity or frequency would produce more current.

20th Century

The discovery of the ionization of gases by ultra-violet light was made by Philipp Lenard in 1900. As the effect was produced across several centimeters of air and made very great positive and small negative ions, it was natural to interpret the phenomenon, as did J. J. Thomson, as a *Hertz effect* upon the solid or liquid particles present in the gas. In 1902, Lenard observed that the energy of individual emitted electrons increased with the frequency (which is related to the color) of the light. This appeared to be at odds with James Clerk Maxwell's wave theory of light, which was thought to predict that the electron energy would be proportional to the intensity of the radiation. In 1902, Lenard observed the variation in electron energy with light frequency. He used a powerful electric arc lamp which enabled him to investigate large changes in intensity, and had sufficient power to enable him to investigate the variation of potential with light frequency. His experiment directly measured potentials, not electron kinetic energy: he found the electron energy by relating it to the maximum stopping potential (voltage) in a phototube. He found that the calculated maximum electron kinetic energy is determined by the frequency of the light.

For example, an increase in frequency results in an increase in the maximum kinetic energy calculated for an electron upon liberation - ultraviolet radiation would require a higher applied stopping potential to stop current in a phototube than blue light. However Lenard's results were qualitative rather than quantitative because of the difficulty in performing the experiments: the experiments needed to be done on freshly cut metal so that the pure metal was observed, but it oxidised in a matter of minutes even in the partial vacuums he used. The current emitted by the surface was determined by the light's intensity, or brightness: doubling the intensity of the light doubled the number of electrons emitted from the surface.

The researches of Langevin and those of Eugene Bloch have shown that the greater part of the Lenard effect is certainly due to this 'Hertz effect'. The Lenard effect upon the gas itself nevertheless does exist. Refound by J. J. Thomson and then more decisively by Palmer, it was studied and showed very different characteristics than those at first attributed to it by Lenard. In 1905, Albert Einstein solved this apparent paradox by describing light as composed of discrete quanta, now called photons, rather than continuous waves. Based upon Max Planck's theory of black-body radiation, Einstein theorized that the energy in each quantum of light was equal to the frequency multiplied by a constant, later called Planck's constant. A photon above a threshold frequency has the required energy to eject a single electron, creating the observed effect. This discovery led to the quantum revolution in physics and earned Einstein the Nobel Prize in Physics in 1921. By wave-particle duality the effect can be analyzed purely in terms of waves though not as conveniently. Albert Einstein's mathematical description of how the photoelectric effect was caused by absorption of quanta of light (now called photons), was in one of his 1905 papers, named "*On a Heuristic Viewpoint Concerning the Production and Transformation of Light*". This paper proposed the simple description of "light quanta", or photons, and showed how they explained such phenomena as the photoelectric effect. His simple explanation in terms of absorption of discrete quanta of light explained the features of the phenomenon and the characteristic frequency. Einstein's explanation of the photoelectric effect won him the Nobel Prize in Physics in 1921. The idea of light quanta began with Max Planck's published law of black-body radiation ("*On the Law of Distribution of Energy in the Normal Spectrum*". Annalen der Physik 4 (1901)) by assuming that Hertzian oscillators could only exist at energies E proportional to the frequency f of the oscillator by $E = hf$, where h is Planck's constant. By assuming that light actually consisted of discrete energy packets, Einstein wrote an equation for the photoelectric effect that agreed with experimental results.

It explained why the energy of photoelectrons were dependent only on the *frequency* of the incident light and not on its *intensity*: a low-intensity, high-frequency source could supply a few high energy photons, whereas a high-intensity, low-frequency source would supply no photons of sufficient individual energy to dislodge any electrons. This was an enormous theoretical leap, but the concept was strongly resisted at first because it contradicted the wave theory of light that followed naturally from James Clerk Maxwell's equations for electromagnetic behaviour, and more generally, the assumption of infinite divisibility of energy in physical systems. Even after experiments showed that Einstein's equations for the photoelectric effect were accurate, resistance to the idea of photons continued, since it appeared to contradict Maxwell's equations, which were well-understood and verified. Einstein's work predicted that the energy of individual ejected electrons increases linearly with the frequency of the light. Perhaps surprisingly, the precise relationship had not at that time been tested.

By 1905 it was known that the energy of photoelectrons increases with increasing *frequency* of incident light and is independent of the *intensity* of the light. However, the manner of the increase was not experimentally determined until 1915 when Robert Andrews Millikan showed that Einstein's prediction was correct. The photoelectric effect helped propel the then-emerging concept of the dualistic nature of light, that light simultaneously possesses the characteristics of both waves and particles, each being manifested according to the circumstances.

The effect was impossible to understand in terms of the classical wave description of light, as the energy of the emitted electrons did not depend on the intensity of the incident radiation. Classical theory predicted that the electrons would 'gather up' energy over a period of time, and then be emitted.

Uses and Effects

Photomultipliers: These are extremely light-sensitive vacuum tubes with a photocathode coated onto part (an end or side) of the inside of the envelope. The photocathode contains combinations of materials such as caesium, rubidium and antimony specially selected to provide a low work function, so when illuminated even by very low levels of light, the photocathode readily releases electrons. By means of a series of electrodes (dynodes) at ever-higher potentials, these electrons are accelerated and substantially increased in number through secondary emission to provide a readily detectable output current. Photomultipliers are still commonly used wherever low levels of light must be detected.

Image Sensors

Video camera tubes in the early days of television used the photoelectric effect, for example, Philo Farnsworth's "Image dissector" used a screen charged by the photoelectric effect to transform an optical image into a scanned electronic signal.

The Gold-Leaf Electroscope

Gold-leaf electroscopes are designed to detect static electricity. Charge placed on the metal cap spreads to the stem and the gold leaf of the electroscope. Because they then have the same charge, the stem and leaf repel each other. This will cause the leaf to bend away from the stem. The electroscope is an important tool in illustrating the photoelectric effect. For example, if the electroscope is negatively charged throughout, there is an excess of electrons and the leaf is separated from the stem. If high-frequency light shines on the cap, the electroscope discharges and the leaf will fall limp. This is because the frequency of the light shining on the cap is above the cap's threshold frequency. The photons in the light have enough energy to liberate electrons from the cap, reducing its negative charge. This will discharge a negatively charged electroscope and further charge a positive electroscope. However, if the electromagnetic radiation hitting the metal cap does not have a high enough frequency (its frequency is below the threshold value for the cap), then the leaf will never discharge, no matter how long one shines the low-frequency light at the cap.

Photoelectron Spectroscopy

Since the energy of the photoelectrons emitted is exactly the energy of the incident photon minus the material's work function or binding energy, the work function of a sample can be determined by bombarding it with a monochromatic X-ray source or UV source, and measuring the kinetic energy distribution of the electrons emitted. Photoelectron spectroscopy is done in a high-vacuum environment, since the electrons would be scattered by gas molecules if they were present. The light source can be a laser, a discharge tube, or a synchrotron radiation source. The concentric hemispherical analyser (CHA) is a typical electron energy analyzer, and uses an electric field to change the directions of incident electrons, depending on their kinetic energies. For every element and core (atomic orbital) there will be a different binding energy. The many electrons created from each of these combinations will show up as spikes in the analyzer output, and these can be used to determine the elemental composition of the sample.

Spacecraft

The photoelectric effect will cause spacecraft exposed to sunlight to develop a positive charge. This can get up to the tens of volts. This can be a major problem, as other parts of the spacecraft in shadow develop a negative charge (up to several kilovolts) from nearby plasma, and the imbalance can discharge through delicate electrical components. The static charge created by the photoelectric effect is self-limiting, though, because a more highly charged object gives up its electrons less easily.

Moon Dust

Light from the sun hitting lunar dust causes it to become charged through the photoelectric effect. The charged dust then repels itself and lifts off the surface of the Moon by electrostatic levitation. This manifests itself almost like an "atmosphere of dust", visible as a thin haze and blurring of distant features, and visible as a dim glow after the sun has set. This was first photographed by the Surveyor programme probes in the 1960s. It is thought that the smallest particles are repelled up to kilometers high, and that the particles move in "fountains" as they charge and discharge.

Night Vision Devices

Photons hitting a thin film of alkali metal or semiconductor material such as gallium arsenide in an image intensifier tube cause the ejection of photoelectrons due to the photoelectric effect. These are accelerated by an electrostatic field where they strike a phosphor coated screen, converting the electrons back int photons. Intensification of the signal is achieved either through acceleration of the electrons or by increasing the number of electrons through secondary emissions, such as with a Micro-channel plate. Sometimes a combination of both methods is used. Additional kinetic energy is required to move an electron out of the conduction band and into the vacuum level. This is known as the electron affinity of the photocathode and is another barrier to photoemission other than the forbidden band, explained by the band gap model. Some materials such as Gallium Arsenide have an effective electron affinity that is below the level of the conduction band. In these materials, electrons that move to the conduction band are all of sufficient energy to be emitted from the material and as such, the film that absorbs photons can be quite thick. These materials are known as negative electron affinity materials.

Cross Section

The photoelectric effect is one interaction mechanism between photons and atoms. It is one of 12 theoretically possible interactions. At the high photon energies comparable to the electron rest energy of 511 keV, Compton scattering, another process, may take place. Above twice this (1.022 MeV) pair production may take place. Compton scattering and pair production are an example of two other competing mechanisms. Indeed, even if the photoelectric effect is the favoured reaction for a particular single-photon bound-electron interaction, the result is also subject to statistical processes and is not guaranteed, albeit the photon has certainly disappeared and a bound electron has been excited (usually K or L shell electrons at nuclear (gamma ray) energies). The probability of the photoelectric effect occurring is measured by the cross section of interaction, ó. This has been found to be a function of the atomic number of the target atom and photon energy.

Bioenergy

Bioenergy is renewable energy made available from materials derived from biological sources. In its most narrow sense it is a synonym to biofuel, which is fuel derived from biological sources. In its broader sense it includes biomass, the biological material used as a biofuel, as well as the social, economic, scientific and technical fields associated with using biological sources for energy.

This is a common misconception, as bioenergy is the energy extracted from the biomass, as the biomass is the fuel and the bioenergy is the energy contained in the fuel. Biomass is any organic material which has stored sunlight in the form of chemical energy. As a fuel it may include wood, wood waste, straw, manure, sugarcane, and many other byproducts from a variety of agricultural processes. There is a slight tendency for the word *bioenergy* to be favoured in Europe compared with *biofuel* in North America.

Solid Biomass

Biomass is material derived from recently living organisms, which includes plants, animals and their byproducts. Manure, garden waste and crop residues are all sources of biomass. It is a renewable energy source based on the carbon cycle, unlike other natural resources such as petroleum, coal, and nuclear fuels. Animal waste is a persistent and unavoidable pollutant produced primarily by the animals housed in industrial-sized farms. There are also agricultural products being grown for biofuel production.

These include corn, and soybeans and to some extent willow and switchgrass on a pre-commercial research level, primarily in the United States; rapeseed, wheat, sugar beet, and willow (15,000 ha in Sweden) primarily in Europe; sugarcane in Brazil; palm oil and miscanthus in Southeast Asia; sorghum and cassava in China; and jatropha in India. Hemp has also been proven to work as a biofuel. Biodegradable outputs from industry, agriculture, forestry and households can be used for biofuel production, using e.g. anaerobic digestion to produce biogas, gasification to produce syngas or by direct combustion. Examples of biodegradable wastes include straw, timber, manure, rice husks, sewage, and food waste. The use of biomass fuels can therefore contribute to waste management as well as fuel security and help to prevent or slow down climate change, although alone they are not a comprehensive solution to these problems.

Electricity Generation from Biomass

Electricity from Sugarcane Bagasse in Brazil: Sucrose accounts for little more than 30% of the chemical energy stored in the mature plant; 35% is in the leaves and stem tips, which are left in the fields during harvest, and 35% are in the fibrous material (bagasse) left over from pressing. The production process of sugar and ethanol in Brazil takes full advantage of the energy stored in sugarcane. Part of the bagasse is currently burned at the mill to provide heat for distillation and electricity to run the machinery. This allows ethanol plants to be energetically self-sufficient and even sell surplus electricity to utilities; current production is 600 MW for self-use and 100 MW for sale. This secondary activity is expected to boom now that utilities have been induced to pay "fair price "(about US$10/GJ or US$0.036/kWh) for 10 year contracts. This is approximately half of what the World Bank considers the reference price for investing in similar projects. The energy is especially valuable to utilities because it is produced mainly in the dry season when hydroelectric dams are running low. Estimates of potential power generation from bagasse range from 1,000 to 9,000 MW, depending on technology. Higher estimates assume gasification of biomass, replacement of current low-pressure steam boilers and turbines by high-pressure ones, and use of harvest trash currently left behind in the fields. For comparison, Brazil's Angra I nuclear plant generates 657 MW. Presently, it is economically viable to extract about 288 MJ of electricity from the residues of one tonne of sugarcane, of which about 180 MJ are used in the plant itself. Thus a medium-size distillery processing 1 million tonnes of sugarcane per year could sell about 5 MW of surplus electricity.

At current prices, it would earn US$ 18 million from sugar and ethanol sales, and about US$ 1 million from surplus electricity sales. With advanced boiler and turbine technology, the electricity yield could be increased to 648 MJ per tonne of sugarcane, but current electricity prices do not justify the necessary investment. (According to one report, the World Bank would only finance investments in bagasse power generation if the price were at least US$19/GJ or US$0.068/kWh.) Bagasse burning is environmentally friendly compared to other fuels like oil and coal. Its ash content is only 2.5% (against 30-50% of coal), and it contains no sulfur. Since it burns at relatively low temperatures, it produces little nitrous oxides. Moreover, bagasse is being sold for use as a fuel (replacing heavy fuel oil) in various industries, including citrus juice concentrate, vegetable oil, ceramics, and tyre recycling. The state of Sγo Paulo alone used 2 million tonnes, saving about US$ 35 million in fuel oil imports. Researchers working with cellulosic ethanol are trying to make the extraction of ethanol from sugarcane bagasse and other plants viable on an industrial scale.

Fossil Fuel

Fossil fuels are fuels formed by natural resources such as anaerobic decomposition of buried dead organisms. The age of the organisms and their resulting fossil fuels is typically millions of years, and sometimes exceeds 650 million years. The fossil fuels, which contain high percentages of carbon, include coal, petroleum, and natural gas. Fossil fuels range from volatile materials with low carbon:hydrogen ratios like methane, to liquid petroleum to nonvolatile materials composed of almost pure carbon, like anthracite coal. Methane can be found in hydrocarbon fields, alone, associated with oil, or in the form of methane clathrates.

It is generally accepted that they formed from the fossilized remains of dead plants and animals by exposure to heat and pressure in the Earth's crust over millions of years. This biogenic theory was first introduced by Georg Agricola in 1556 and later by Mikhail Lomonosov in the 18th century. It was estimated by the Energy Information Administration that in 2007 primary sources of energy consisted of petroleum 36.0%, coal 27.4%, natural gas 23.0%, amounting to an 86.4% share for fossil fuels in primary energy consumption in the world. Non-fossil sources in 2006 included hydroelectric 6.3%, nuclear 8.5%, and others (geothermal, solar, tide, wind, wood, waste) amounting to 0.9 percent.

World energy consumption was growing about 2.3% per year. Fossil fuels are non-renewable resources because they take millions of years to form, and reserves are being depleted much faster than new ones are being made. The production and use of fossil fuels raise environmental concerns. A global movement toward the generation of renewable energy is therefore under way to help meet increased energy needs. The burning of fossil fuels produces around 21.3 billion tonnes (21.3 gigatonnes) of carbon dioxide (CO_2) per year, but it is estimated that natural processes can only absorb about half of that amount, so there is a net increase of 10.65 billion tonnes of atmospheric carbon dioxide per year (one tonne of atmospheric carbon is equivalent to 44/12 or 3.7 tonnes of carbon dioxide). Carbon dioxide is one of the greenhouse gases that enhances radiative forcing and contributes to global warming, causing the average surface temperature of the Earth to rise in response, which most climate scientists agree will cause major adverse effects.

Origin

Petroleum and natural gas are formed by the anaerobic decomposition of remains of organisms including phytoplankton and zooplankton that settled to the sea (or lake) bottom in large quantities under anoxic conditions, millions of years ago. Over geological time, this organic matter, mixed with mud, got buried under heavy layers of sediment. The resulting high levels of heat and pressure caused the organic matter to chemically alter, first into a waxy material known as kerogen which is found in oil shales, and then with more heat into liquid and gaseous hydrocarbons in a process known as catagenesis.

There is a wide range of organic, or hydrocarbon, compounds in any given fuel mixture. The specific mixture of hydrocarbons gives a fuel its characteristic properties, such as boiling point, melting point, density, viscosity, etc. Some fuels like natural gas, for instance, contain only very low boiling, gaseous components. Others such as gasoline or diesel contain much higher boiling components. Terrestrial plants, on the other hand, tend to form coal and methane. Many of the coal fields date to the Carboniferous period of Earth's history. Terrestrial plants also form type III kerogen, a source of natural gas.

Importance

Fossil fuels are of great importance because they can be burned (oxidized to carbon dioxide and water), producing significant amounts of energy per unit weight. The use of coal as a fuel predates recorded history. Coal was used to run furnaces for the melting of metal ore. Semi-solid hydrocarbons from seeps were also burned in ancient times, but these materials were mostly used for waterproofing and embalming. Commercial exploitation of petroleum, largely as a replacement for oils from animal sources (notably whale oil), for use in oil lamps began in the 19th century. Natural gas, once flared-off as an unneeded byproduct of petroleum production, is now considered a very valuable resource. Heavy crude oil, which is much more viscous than conventional crude oil, and tar sands, where bitumen is found mixed with sand and clay, are becoming more important as sources of fossil fuel. Oil shale and similar materials are sedimentary rocks containing kerogen, a complex mixture of high-molecular weight organic compounds, which yield synthetic crude oil when heated (pyrolyzed). These materials have yet to be exploited commercially. These fuels can be employed in internal combustion engines, fossil fuel power stations and other uses. Prior to the latter half of the 18th century, windmills and watermills provided the energy needed for industry such as milling flour, sawing wood or pumping water, and burning wood or peat provided domestic heat. The widescale use of fossil fuels, coal at first and petroleum later, to fire steam engines enabled the Industrial Revolution. At the same time, gas lights using natural gas or coal gas were coming into wide use. The invention of the internal combustion engine and its use in automobiles and trucks greatly increased the demand for gasoline and diesel oil, both made from fossil fuels. Other forms of transportation, railways and aircraft, also required fossil fuels. The other major use for fossil fuels is in generating electricity and as feedstock for the petrochemical industry. Tar, a leftover of petroleum extraction, is used in construction of roads.

Levels and Flows

Levels of primary energy sources are the reserves in the ground. Flows are production. The most important part of primary energy sources are the carbon based fossil energy sources. Coal, oil, and natural gas provided 79.6% of primary energy production during 2002 (in million tonnes of oil equivalent (mtoe)) (34.9+23.5+21.2).

Levels (proved reserves) during 2005-2007

- Coal: 997,748 million short tonnes (905 billion metric tonnes), 4,416 billion barrels (702.1 km^3) of oil equivalent
- Oil: 1,119 billion barrels (177.9 km^3) to 1,317 billion barrels (209.4 km^3)
- Natural gas: 6,183-6,381 trillion cubic feet (175-181 trillion cubic metres), 1,161 billion barrels ($184.6\times10^{\triangle 9}$ m^3) of oil equivalent

Flows (daily production) during 2006

- Coal: 18,476,127 short tonnes (16,761,260 metric tonnes), 52,000,000 barrels (8,300,000 m^3) of oil equivalent per day
- Oil: 84,000,000 barrels per day (13,400,000 m^3/d)
- Natural gas: 104,435 billion cubic feet (2,960 billion cubic metres), 19,000,000 barrels (3,000,000 m^3) of oil equivalent per day

Years of production left in the ground with the current proved reserves and flows above

- Coal: 148 years
- Oil: 43 years
- Natural gas: 61 years

Years of production left in the ground with the most optimistic proved reserve estimates (Oil & Gas Journal, World Oil)

- Coal: 417 years
- Oil: 43 years
- Natural gas: 167 years

The calculation above assumes that the product could be produced at a constant level for that number of years and that all of the proved reserves could be recovered. In reality, consumption of all three resources has been increasing. While this suggests that the resource will be used up more quickly, in reality, the production curve is much more akin to a bell curve. At some point in time, the production of each resource within an area, country, or globally will reach a maximum value, after which, the production will decline until it reaches a point where is no longer economically feasible or physically possible to produce. See Hubbert peak theory for detail on this decline curve with regard to petroleum. Note also that proved reserve estimates do not include strategic reserves, which (globally) amount to 4.1 billion more barrels. The above discussion emphasizes worldwide energy balance. It is also valuable to understand the ratio of reserves to annual consumption (R/C) by region or country. For example, energy policy of the United Kingdom recognizes that Europe's R/C value is 3.0, very low by world standards, and exposes that region to energy vulnerability. Alternatives to fossil fuels are a subject of intense debate worldwide.

Limits and Alternatives

The principle of supply and demand suggests that as hydrocarbon supplies diminish, prices will rise. Therefore higher prices will lead to increased alternative, renewable energy supplies as previously uneconomic sources become sufficiently economical to exploit. Artificial gasolines and other renewable energy sources currently require more expensive production and processing technologies than conventional petroleum reserves, but may become economically viable in the near future. See Energy development. Different alternative sources of energy include nuclear, hydroelectric, solar, wind, and geothermal.

Environmental Effects

In the United States, more than 90% of greenhouse gas emissions come from the combustion of fossil fuels. Combustion of fossil fuels also produces other air pollutants, such as nitrogen oxides, sulfur dioxide, volatile organic compounds and heavy metals.

According to Environment Canada:

> *"The electricity sector is unique among industrial sectors in its very large contribution to emissions associated with nearly all air issues. Electricity generation produces a large share of Canadian nitrogen oxides and sulphur dioxide emissions, which contribute to smog and acid rain and the formation of fine particulate matter. It is the largest uncontrolled industrial source of mercury emissions in Canada. Fossil fuel-fired electric power plants also emit carbon dioxide, which may contribute to climate change. In addition, the sector has significant impacts on water and habitat and species. In particular, hydro dams and transmission lines have significant effects on water and biodiversity." According to U.S.*

Scientist Jerry Mahlman and USA Today: Mahlman, who crafted the IPCC language used to define levels of scientific certainty, says the new report will lay the blame at the feet of fossil fuels with "virtual certainty," meaning 99% sure. That's a significant jump from "likely," or 66% sure, in the group's last report in 2001, Mahlman says. His role in this year's effort involved spending two months reviewing the more than 1,600 pages of research that went into the new assessment. Combustion of fossil fuels generates sulfuric, carbonic, and nitric acids, which fall to Earth as acid rain, impacting both natural areas and the built environment. Monuments and sculptures made from marble and limestone are particularly vulnerable, as the acids dissolve calcium carbonate. Fossil fuels also contain radioactive materials, mainly uranium and thorium, which are released into the atmosphere. In 2000, about 12,000 tonnes of thorium and 5,000 tonnes of uranium were released worldwide from burning coal. It is estimated that during 1982, US coal burning released 155 times as much radioactivity into the atmosphere as the Three Mile Island incident. However, this radioactivity from coal burning is minuscule at each source and has not shown to have any adverse effect on human physiology. Burning coal also generates large amounts of bottom ash and fly ash. These materials are used in a wide variety of applications, utilizing, for example, about 40% of the US production.

Harvesting, processing, and distributing fossil fuels can also create environmental concerns. Coal mining methods, particularly mountaintop removal and strip mining, have negative environmental impacts, and offshore oil drilling poses a hazard to aquatic organisms. Oil refineries also have negative environmental impacts, including air and water pollution. Transportation of coal requires the use of diesel-powered locomotives, while crude oil is typically transported by tanker ships, each of which requires the combustion of additional fossil fuels. Environmental regulation uses a variety of approaches to limit these emissions, such as command-and-control (which mandates the amount of pollution or the technology used), economic incentives, or voluntary programmes.

An example of such regulation in the USA is the "EA is implementing policies to reduce airborne mercury emissions. Under regulations issued in 2005, coal-fired power plants will need to reduce their emissions by 70 percent by 2018.". In economic terms, pollution from fossil fuels is regarded as a negative externality. Taxation is considered one way to make societal costs explicit, in order to 'internalize' the cost of pollution. This aims to make fossil fuels more expensive, thereby reducing their use and the amount of pollution associated with them, along with raising the funds necessary to counteract these factors. Former CIA Director James Woolsey recently outlined the national security arguments in favour of moving away from fossil fuels.

Global Dimming

Global dimming is the gradual reduction in the amount of global direct irradiance at the Earth's surface that was observed for several decades after the start of systematic measurements in the 1950s. The effect varies by location, but worldwide it has been estimated to be of the order of a 4% reduction over the three decades from 1960–1990. However, after discounting an anomaly caused by the eruption of Mount Pinatubo in 1991, a very slight reversal in the overall trend has been observed.

It is thought to have been caused by an increase in particulates such as sulfate aerosols in the atmosphere due to human action. Global dimming has interfered with the hydrological cycle by reducing evaporation and may have reduced rainfall in some areas. Global dimming also creates a cooling effect that may have partially masked the effect of greenhouse gases on global warming. Deliberate manipulation of this dimming effect is now being considered as a geoengineering technique to reduce the impact of global warming.

Causes and Effects

It is thought that global dimming is probably due to the increased presence of aerosol particles in the atmosphere caused by human action. Aerosols and other particulates absorb solar energy and reflect sunlight back into space. The pollutants can also become nuclei for cloud droplets. Water droplets in clouds coalesce around the particles. Increased pollution causes more particulates and thereby creates clouds consisting of a greater number of smaller droplets that is, the same amount of water is spread over more droplets).

The smaller droplets make clouds more reflective, so that more incoming sunlight is reflected back into space and less reaches the Earth's surface. In models, these smaller droplets also decrease rainfall. Clouds intercept both heat from the sun and heat radiated from the

Earth. Their effects are complex and vary in time, location, and altitude. Usually during the daytime the interception of sunlight predominates, giving a cooling effect; however, at night the re-radiation of heat to the Earth slows the Earth's heat loss.

Research

In the late-1960s, Mikhail Ivanovich Budyko worked with simple two-dimensional energy-balance climate models to investigate the reflectivity of ice. He found that the ice-albedo feedback created a positive feedback loop in the Earth's climate system. The more snow and ice, the more solar radiation is reflected bac into space and hence the colder Earth grows and the more it snows. Other studies found that pollution or a volcano eruption could provoke the onset of an ice age. In the mid-1980s, Atsumu Ohmura, a geography researcher at the Swiss Federal Institute of Technology, found that solar radiation striking the Earth's surface had declined by more than 10% over the three previous decades. His findings appeared to contradict global warming—the global temperature had been generally rising since the 70s. Less light reaching the earth seemed to mean that it should cool. Ohmura published his findings "Secular variation of global radiation in Europe" in 1989. This was soon followed by others: Viivi Russak in 1990 "Trends of solar radiation, cloudiness and atmospheric transparency during recent decades in Estonia", and Beate Liepert in 1994 "Solar radiation in Germany — Observed trends and an assessment of their causes". Dimming has also been observed in sites all over the former Soviet Union. Gerry Stanhill who studied these declines worldwide in many papers coined the term "global dimming". Independent research in Israel and the Netherlands in the late 1980s showed an apparent reduction in the amount of sunlight, despite widespread evidence that the climate was actually becoming hotter. The rate of dimming varies around the world but is on average estimated at around 2–3% per decade. The trend reversed in the early 1990s. It is difficult to make a precise measurement, due to the difficulty in accurately calibrating the instruments used, and the problem of spatial coverage. Nonetheless, the effect is almost certainly present. The effect (2–3%, as above) is due to changes within the Earth's atmosphere; the value of the solar radiation at the top of the atmosphere has not changed by more than a fraction of this amount. The effect varies greatly over the planet, but estimates of the terrestrial surface average value are:

- 5.3% (9 W/m^2); over 1958–85 (Stanhill and Moreshet, 1992)
- 2%/decade over 1964–93 (Gilgen *et al.*, 1998)

- 2.7%/decade (total 20 W/m^2); up to 2000 (Stanhill and Cohen, 2001)
- 4% over 1961–90 (Liepert 2002)

Note that these numbers are for the terrestrial surface and not really a global average. Whether dimming (or brightening) occurred over the ocean has been a bit of an unknown though a specific measurement measured effects some 400 miles (643.7 km) from India over the Indian Ocean towards the Maldives Islands. Regional effects probably dominate but are not strictly confined to the land area, and the effects will be driven by regional air circulation. A 2009 review by Wild et al. found that widespread variation in regional and time effects. There was solar brightening beyond 2000 at numerous stations in Europe, the United States, and Korea. The brightening seen at sites in Antarctica during the 1990s, influenced by recovering from the Mount Pinatubo volcanic eruption in 1991, fades after 2000. The brightening tendency also seems to level off at sites in Japan. In China there is some indication for a renewed dimming, after the stabilization in the 1990s. A continuation of the long-lasting dimming is also noted at the sites in India. Overall, the available data suggest continuation of the brightening beyond the year 2000 at numerous locations, yet less pronounced and coherent than during the 1990s, with more regions with no clear changes or declines. Therefore, globally, greenhouse warming after 2000 may be less modulated by surface solar variations than in prior decades. The largest reductions are found in the northern hemisphere mid-latitudes. Visible light and infrared radiation seem to be most affected rather than the ultraviolet part of the spectrum.

Pan Evaporation Data

Over the last 50 or so years, pan evaporation has been carefully monitored. For decades, nobody took much notice of the pan evaporation measurements. But in the 1990s in Europe, Israel, and North America, scientists spotted something that at the time was considered very strange: the rate of evaporation was falling although they had expected it to increase due to global warming.

The same trend has been observed in China over a similar period. A decrease in solar irradiance is cited as the driving force. However, unlike in other areas of the world, in China the decrease in solar irradiance was not always accompanied by an increase in cloud cover and precipitation. It is believed that aerosols may play a critical role in the decrease of solar irradiance in China. BBC Horizon producer David Sington believes that many climate scientists regard the pan-evaporation data as the most convincing evidence of solar dimming. Pan evaporation experiments are easy to reproduce with low-cost equipment, there are many pans used for agriculture all over the world and in many instances the data has been collected for nearly a half century. However, pan evaporation depends on some additional factors besides net radiation from the sun. The other two major factors are vapor pressure deficit and wind speed. The ambient temperature turns out to be a negligible factor. The pan evaporation data corroborates the data gathered by radiometer and fills in the gaps in the data obtained using pyranometers. With adjustments to these factors, pan evaporation data has been compared to results of climate simulations.

Probable Causes

The incomplete combustion of fossil fuels (such as diesel) and wood releases black carbon into the air. Though black carbon, most of which is soot, is an extremely small component of air pollution at land surface levels, the phenomenon has a significant heating effect on the atmosphere at altitudes above two kilometers (6,562 ft). Also, it dims the surface of the ocean by absorbing solar radiation. Experiments in the Maldives (comparing the atmosphere over the northern and southern islands) in the 1990s showed that the effect of macroscopic pollutants in the atmosphere at that time (blown south from India) caused about a 10% reduction in sunlight reaching the surface in the area under the pollution cloud — a much greater reduction than expected from the presence of the particles themselves.

Prior to the research being undertaken, predictions were of a 0.5–1% effect from particulate matter; the variation from prediction may be explained by cloud formation with the particles acting as the focus for droplet creation. Clouds are very effective at reflecting light back out into space. The phenomenon underlying global dimming may also have regional effects. While most of the earth has warmed, the regions that are downwind from major sources of air pollution (specifically sulfur dioxide emissions) have generally cooled. This may explain the cooling of the eastern United States relative to the warming western part.

However some research shows that black carbon will actually increase global warming, being second only to CO_2. They believe that soot will absorb solar energy and transport it to other areas such as the Himalayas where glacial melting occurs. It can also darken Arctic ice reducing reflectivity and increasing absorption of solar radiation. Some climate scientists have theorized that aircraft contrails (also called vapor trails) are implicated in global dimming, but the constant flow of air traffic previously meant that this could not be tested.

The near-total shutdown of civil air traffic during the three days following the September 11, 2001 attacks afforded a unique opportunity in which to observe the climate of the United States absent from the effect of contrails. During this period, an increase in diurnal temperature variation of over 1 °C (1.8 °F) was observed in some parts of the U.S., i.e. aircraft contrails may have been raising nighttime temperatures and/or lowering daytime temperatures by much more than previously thought. Airborne volcanic ash can reflect the Sun's rays back into space and thereby contribute to cooling the planet. Dips in earth temperatures have been observed after large volcano eruptions such as Mount Agung in Bali that erupted in 1963, El Chichon (Mexico) 1983, Ruiz (Colombia) 1985, and Pinatubo (Philippines) 1991. But even for major eruptions, the ash clouds remain only for relatively short periods.

Recent Reversal of the Trend

Wild *et al.*, using measurements over land, report brightening since 1990, and Pinker *et al.* found that slight dimming continued over land while brightening occurred over the ocean. Hence, over the land surface, Wild *et al.* and Pinker *et al.* disagree. A 2007 NASA sponsored satellite-based study sheds light on the puzzling observations by other scientists that the amount of sunlight reaching Earth's surface had been steadily declining in recent decades, began to reverse around 1990. This switch from a "global dimming" trend to a "brightening" trend happened just as global aerosol levels started to decline. It is likely that at least some of this change, particularly over Europe, is due to decreases in airborne pollution.

Most governments of developed nations have taken steps to reduce aerosols released into the atmosphere, which helps reduce global dimming. Sulfate aerosols have declined significantly since 1970 with the Clean Air Act in the United States and similar policies in Europe. The Clean Air Act was strengthened in 1977 and 1990. According to the EPA, from 1970 to 2005, total emissions of the six principal air pollutants, including PM's, dropped by 53% in the US. In 1975, the masked effects of trapped greenhouse gases finally started to emerge and have dominated ever since. The Baseline Surface Radiation Network(BSRN) has been collecting surface measurements. BSRN was started in the early 1990s and updated the archives in this time. Analysis of recent data reveals that the surface of the planet has brightened by about 4% in the past decade. The brightening trend is corroborated by other data, including satellite analyses.

Relationship to Hydrological Cycle

Pollution produced by humans may be seriously weakening the Earth's water cycle — reducing rainfall and threatening fresh water supplies. A 2001 study by researchers at the Scripps Institution of Oceanography suggests that tiny particles of soot and other pollutants have a significant effect on the hydrological cycle. According to Veerabhadran Ramanathan, "the energy for the hydrological cycle comes from sunlight. As sunlight heats the ocean, water escapes into the atmosphere and falls out as rain. So as aerosols cut down sunlight by large amounts, they may be spinning down the hydrological cycle of the planet." Large scale changes in weather patterns may also have been caused by global dimming. Climate modelers speculatively suggest that this reduction in solar radiation at the surface may have led to the failure of the monsoon in sub-Saharan Africa during the 1970s and 1980s, together with the associated famines such as the Sahel drought, caused by Northern hemisphere pollution cooling the Atlantic. Because of this, the Tropical rain belt may not have risen to its northern latitudes, thus causing an absence of seasonal rains. This claim is not universally accepted and is very difficult to test. However a 2009 Chinese study of 50 years of continuous data found that though most parts of eastern China saw no significant change in the amount of water held by the atmosphere but light rains had decreased. In addition, where the atmosphere transported water vapor didn't coincide with light rain frequency.

The researchers then modeled the effect of aerosols and also concluded the overall effect was that water drops in polluted cases are up to 50 percent smaller than in pristine skies. They concluded smaller size impedes the formation of rain clouds and the falling of light rain beneficial for agriculture; a different effect than reducing solar irradiance, but still a direct result from the presence of aerosols. The 2001 study by researchers at the Scripps Institution of Oceanography concluded that the imbalance between global dimming and global warming at the surface leads to weaker turbulent heat fluxes to the atmosphere. This means globally reduced evaporation and hence precipitation occur in a dimmer and warmer world, which could ultimately lead to a more humid atmosphere in which it rains less. A natural form of large scale environmental shading/dimming has been identified that affected the 2006 northern hemisphere hurricane season. The NASA study found that several major dust storms in June and

July in the Sahara desert sent dust drifting over the Atlantic Ocean and through several effects caused cooling of the waters — and thus dampening the development of hurricanes.

Relationship to Global Warming

Some scientists now consider that the effects of global dimming have masked the effect of global warming to some extent and that resolving global dimming may therefore lead to increases in predictions of future temperature rise. According to Beate Liepert, "We lived in a global warming plus a global dimming world and now we are taking out global dimming.

So we end up with the global warming world, which will be much worse than we thought it will be, much hotter." The magnitude of this masking effect is one of the central problems in climate change with significant implications for future climate changes and policy responses to global warming. Interactions between the two theories for climate modification have also been studied, as global warming and global dimming are neither mutually exclusive nor contradictory. In a paper published on March 8, 2005 in the American Geophysical Union's Geophysical Research Letters, a research team led by Anastasia Romanou of Columbia University's Department of Applied Physics and Mathematics, New York, also showed that the apparently opposing forces of global warming and global dimming can occur at the same time.

Global dimming interacts with global warming by blocking sunlight that would otherwise cause evaporation and the particulates bind to water droplets. Water vapor is the major greenhouse gas. On the other hand, global dimming is affected by evaporation and rain. Rain has the effect of clearing out polluted skies.

Brown clouds have been found to amplify global warming according to Veerabhadran Ramanathan, an atmospheric chemist at the Scripps Institution of Oceanography in La Jolla, CA. "The conventional thinking is that brown clouds have masked as much as 50 percent of global warming by greenhouse gases through so-called global dimming... While this is true globally, this study reveals that over southern and eastern Asia, the soot particles in the brown clouds are in fact amplifying the atmospheric warming trend caused by greenhouse gases by as much as 50 percent."

Possible Use to Mitigate Global Warming

Some scientists have suggested using aerosols to stave off the effects of global warming as an emergency geoengineering measure. In 1974, Mikhail Budyko suggested that if global warming became a problem, the planet could be cooled by burning sulfur in the stratosphere, which would create a haze. An increase in planetary albedo of just 0.5 percent is sufficient to halve the effect of a CO_2 doubling. The simplest solution would be to simply emit more sulfates, which would end up in troposphere - the lowest part of the atmosphere. If this was done, Earth would still face many problems, such as:

- Using sulfates causes environmental problems such as acid rain
- Using carbon black causes human health problems
- Dimming causes ecological problems such as changes in evaporation and rainfall patterns
- Droughts and/or increased rainfall cause problems for agriculture
- Aerosol has a relatively short lifetime

The solution actually advocated is transporting sulfates into the next higher layer of the atmosphere - stratosphere. Aerosols in the stratosphere last years instead of weeks - so only a relatively smaller (though still large) amount of sulfate emissions would be necessary, and side effects would be less.

This would require developing an efficient way to transport large amounts of gases into stratosphere, many of which have been proposed though none are known to be effective or economically viable. In a blog post, Gavin Schmidt stated that "Ideas that we should increase aerosol emissions to counteract global warming have been described as a 'Faustian bargain' because that would imply an ever increasing amount of emissions in order to match the accumulated greenhouse gas in the atmosphere, with ever increasing monetary and health costs."

Nuclear Fuel

Nuclear fuel is a material that can be 'consumed' by fission or fusion to derive nuclear energy, by analogy to chemical fuel that is burned for energy. Nuclear fuels are the most dense sources of energy available. *Nuclear fuel* in a nuclear fuel cycle can refer to the fuel itself, or to physical objects (for example bundles composed of fuel rods) composed of the fuel material, mixed with structural, neutron moderating, or neutron reflecting materials.

Most nuclear fuels contain heavy fissile elements that can be made to undergo a nuclear fission chain reaction in a nuclear reactor. The most common fissile nuclear fuels are Uranium 235 (^{235}U) and Plutonium 239 (^{239}Pu). The actions of mining, refining, purifying, using, and ultimately disposing of these elements together make up the nuclear fuel cycle. Not all nuclear fuels are used in fission reactors. Plutonium-238 and some other elements are used to produce small amounts of nuclear power by radioactive decay in radioisotope thermoelectric generators and other atomic batteries. Light nuclides such as ^3H (tritium) are used as fuel for nuclear fusion.

Oxide Fuel

For fission reactors, the fuel (typically based on Uranium) is usually based on the metal oxide; the oxides are used rather than the metals themselves because the oxide melting point is much higher than that of the metal and because it cannot burn, being already in the oxidized state. The thermal conductivity of uranium dioxide is low; it is affected by porosity and burn-up. The burn-up results in fission products being dissolved in the lattice (such as lanthanides), the precipitation of fission products such as palladium, the formation of fission gas bubbles due to fission products such as xenon and krypton and radiation damage of the lattice. The low thermal conductivity can lead to overheating of the center part of the pellets during use. The porosity results in a decrease in both the thermal conductivity of the fuel and the swelling which occurs during use. According to the International Nuclear Safety Center the thermal conductivity of uranium dioxide can be predicted under different conditions by a series of equations. The bulk density of the fuel can be related to the thermal conductivity

Where ρ is the bulk density of the fuel and ρ_{td} is the theoretical density of the uranium dioxide.

Then the thermal conductivity of the porous phase (K_f) is related to the conductivity of the perfect phase (K_o, no porosity) by the following equation. Note that s is a term for the shape factor of the holes.

$$K_f = K_o(1 \text{ ” } p/1 + (s \text{ ” } 1)p)$$

Rather than measuring the thermal conductivity using the traditional methods in physics such as Lees' disk, the Forbes' method or Searle's bar it is common to use a laser flash method where a small disc of fuel is placed in a furnace.

UOX

Uranium dioxide is a black semiconductor solid. It can be made by reacting uranyl nitrate with a base (ammonia) to form a solid (ammonium uranate). It is heated (calcined) to form U_3O_8 that can then be converted by heating in an argon / hydrogen mixture (700 °C) to form UO_2. The UO_2 is then mixed with an organic binder and pressed into pellets, these pellets are then fired at a much higher temperature (in H_2/Ar) to sinter the solid. The aim is to form a dense solid which has few pores. The thermal conductivity of uranium dioxide is very low compared with that of zirconium metal, and it goes down as the temperature goes up. It is important to note that the corrosion of uranium dioxide in an aqueous environment is controlled by similar electrochemical processes to the galvanic corrosion of a metal surface.

MOX

Mixed oxide, or MOX fuel, is a blend of plutonium and natural or depleted uranium which behaves similarly (though not identically) to the enriched uranium feed for which most nuclear reactors were designed. MOX fuel is an alternative to low enriched uranium (LEU) fuel used in the light water reactors which predominate nuclear power generation. Some concern has been expressed that used MOX cores will introduce new disposal challenges, though MOX is itself a means to dispose of surplus plutonium by transmutation. Currently (March, 2005) reprocessing of commercial nuclear fuel to make MOX is done in England and France, and to a lesser extent in Russia, India and Japan. China plans to develop fast breeder reactors and reprocessing. The Global Nuclear

Energy Partnership, is a U.S. plan to form an international partnership to see spent nuclear fuel reprocessed in a way that renders the plutonium in it usable for nuclear fuel but not for nuclear weapons. Reprocessing of spent commercial-reactor nuclear fuel has not been permitted in the United States due to nonproliferation considerations. All of the other reprocessing nations have long had nuclear weapons from military-focused "research"-reactor fuels except for Japan.

Metal Fuel

Metal fuels have the advantage of a much higher heat conductivity than oxide fuels but cannot survive equally high temperatures. Metal fuels have a long history of use, stretching from the Clementine reactor in 1946 to many test and research reactors. Metal fuels have the potential for the highest fissile atom density. Metal fuels are normally alloyed, but some metal fuels have been made with pure uranium metal. Uranium alloys that have been used include uranium aluminum, uranium zirconium, uranium silicon, uranium molybdenum, and uranium zirconium hydride.

Any of the aforementioned fuels can be made with plutonium and other actinides as part of a closed nuclear fuel cycle. Metal fuels have been used in water reactors and liquid metal fast breeder reactors, such as EBR-II.

TRIGA Fuel

TRIGA fuel is used in TRIGA (Training, Research, Isotopes, General Atomics) reactors. The TRIGA reactor uses uranium-zirconium-hydride (UZrH) fuel, which has a prompt negative temperature coefficient, meaning that as the temperature of the core increases, the reactivity decreases—so it is highly unlikely for a meltdown to occur.

Most cores that use this fuel are "high leakage" cores where the excess leaked neutrons can be utilized for research. TRIGA fuel was originally designed to use highly enriched uranium, however in 1978 the U.S. Department of Energy launched its Reduced Enrichment for Research Test Reactors programme, which promoted reactor conversion to low-enriched uranium fuel.

A total of 35 TRIGA reactors have been installed at locations across the USA. A further 35 reactors have been installed in other countries.

Actinide Fuel

In a fast neutron reactor, the minor actinides produced by neutron capture of uranium and plutonium can be used as fuel. Metal actinide fuel is typically an alloy of zirconium, uranium, plutonium and the minor actinides. It can be made inherently safe as thermal expansion of the metal alloy will increase neutron leakage.

Ceramic Fuels

Ceramic fuels othr than oxides have the advantage of high heat conductivities and melting points, but they are more prone to swelling than oxide fuels and are much less well understood.

Uranium Nitride

This is often the fuel of choice for reactor designs that NASA produces, one advantage is that UN has a better thermal conductivity than UO_2. Uranium nitride has a very high melting point. This fuel has the disadvantage that unless ^{15}N was used (in place of the more common ^{14}N) that a large amount of ^{14}C would be generated from the nitrogen by the (n,p) reaction.

As the nitrogen required for such a fuel would be so expensive it is likely that the fuel would have to be reprocessed by a *pyro* method to enable to the ^{15}N to be recovered.

It is likely that if the fuel was processed and dissolved in nitric acid that the nitrogen enriched with ^{15}N would be dilted with the common ^{14}N.

Uranium Carbide

Much of what is known about uranium carbide is in the form of pin-type fuel elements for liquid metal fast breeder reactors during their intense study during the '60s and '70s. However, recently there has been a revived interest in uranium carbide in the form of plate fuel and most notably, mico fuel particles (such as TRISO particles). The high thermal conductivity and high melting point make uranium carbide an attractive fuel. In addition, because of the absence of oxygen in this fuel (durng the course of irradiation, excess gas pressure can build from the formation of O_2 or other gases) as well as the ability to complement a ceramic coating (a ceramic-ceramic interface has structural and chemical advantages), uranium carbide could be the ideal fuel candidate for certain Generation IV reactors such as the gas-cooled fast reactor.

Liquid Fuels

Molten Salts: These include fuels where the fuel is dissolved in the coolant. They were used in the Molten Salt Reactor Experiment and numerous other liquid core reactor experiments, such as the liquid fluoride thorium reactor. The liquid fuel for the molten salt reactor was a mixture of lithium, beryllium, thorium and uranium fluorides: $LF\text{-}BeF_2\text{-}ThF_4\text{-}UF_4$ (72-16-12-0.4 mol%). It had a peak operating temperature of 705 °C in the experiment, but could have operated at much higher temperatures, since the boiling point of the molten salt was in excess of 1400 °C.

Aqueous Solutions of Uranyl Salts

The aqueous homogeneous reactors use a solution of uranyl sulfate or other uranium salt in water. This homogenous reactor type has not been used for any large power reactors. One of its disadvantages is that the fuel is in a form which is easy to disperse in the event of an accident.

Common Physical Forms of Nuclear Fuel

Uranium dioxide (UO_2) powder is compacted to cylindrical pellets and sintered at high temperatures to produce ceramic nuclear fuel pellets with a high density and well defined physical properties and chemical composition. A grinding process is used to achieve a uniform cylindrica geometry wit narrow tolerances. Such fuel pellets are then stacked and filled into the metallic tubes. The metal used for the tubes depends on the design of the reactor. Stainless steel was used in the past, but most reactors now use a zirconium alloy which, in addition to being highly corrosion-resistant, has low neutron absorption. The tubes containing the fuel pellets are sealed: these tubes are called fuel rods. The finished fuel rods are grouped into fuel assemblies that are used to build up the core of a power reactor.

Cladding is the outer layer of the fuel rods, standing between the coolant and the nuclear fuel. It is made of a corrosion-resistant material with low absorption cross section for thermal neutrons, usually Zircaloy or steel in modern constructions, or magnesium with small amount of aluminium and other metals for the now-obsolete Magnox reactors. Cladding prevents radioactive fission fragments from esaping the fuel into the coolant and contaminating it.

PWR Fuel

Pressurized water reactor (PWR) fuel consists of cylindrical rods put into bundles. A uranium oxide ceramic is formed into pellets and inserted into Zircaloy tubes that are bundled together. The Zircaloy tubes are about 1 cm in diameter, and the fuel cladding gap is filled with helium gas to improve the conduction of heat from the fuel to the cladding. There are about 179-264 fuel rods per fuel bundle and about 121 to 193 fuel bundles are loaded into a reactor core. Generally, the fuel bundles consist of fuel rods bundled 14x14 to 17x17.

PWR fuel bundles are about 4 meters in length. In PWR fuel bundles, control rods are inserted through the top directly into the fuel bundle. The fuel bundles usually are enriched several percent in ^{235}U. The uranium oxide is dried before inserting into the tubes to try to eliminate moisture in the ceramic fuel that can lead to corrosion and hydrogen embrittlement. The Zircaloy tubes are pressurized with helium to try to minimize pellet-cladding interaction which can lead to fuel rod failure over long periods.

BWR Fuel

In boiling water reactors (BWR), the fuel is similar to PWR fuel except that the bundles are "canned"; that is, there is a thin tube surrouning each bundle. This is primarily done to prevent local density variations from affecting neutronics and thermal hydraulics of the reactor core. In modern BWR fuel bundles, there are either 91, 92, or 96 fuel rods per assembly depending on the manufacturer.

A range between 368 assemblies for the smallest and 800 assemblies for the largest U.S. BWR forms the reactor core. Each BWR fuel rod is back filled with helium to a pressure of about three atmospheres (300 kPa).

CANDU Fuel

CANDU fuel bundles are about a half meter in length and 10 cm in diameter. They consist of sintered (UO_2) pellets in zirconium alloy tubes, welded to zirconium alloy end plates. Each bundle is roughly 20 kg, and a typical core loading is on the order of 4500-6500 bundles, depending on the design. Modern types typically have 37 identical fuel pins radially arranged about the long axis of the bundle, but in the past several different configurations and numbers of pins have been used. The CANFLEX bundle has 43 fuel elements, with two element sizes. It is also about 10 cm (4 inches) in diameter, 0.5 m (20 in) long and weighs about 20 kg (44 lb) and replaces the 37-pin standard bundle. It has been designed specifically to increase fuel performance by utilizing two different pin diameters. Current CANDU designs do not need enriched uranium to achieve criticality (due to their more efficient heavy water moderator), however, some newer concepts call for low enrichment to help reduce the size of the reactors.

Less Common Fuel Forms

Various other nuclear fuel forms find use in specific applications, but lack the widespread use of those found in BWRs, PWRs, and CANDU power plants. Many of these fuel forms are only found in research reactors, or have military applications.

Magnox Fuel

Magnox reactors are pressurised, carbon dioxide cooled, graphite moderated reactors using natural uranium (i.e. unenriched) as fuel and magnox alloy as fuel cladding. Working pressure varies from 6.9 to 19.35 bar for the steel pressure vessels, and the two reinforced concrete designs operated at 24.8 and 27 bar. Magnox is also the name of an alloy—mainly of magnesium with small amounts of aluminium and other metals—used in cladding unenriched uranium metal fuel with a non-oxidising covering to contain fission products. Magnox is short for Magnesium non-oxidising. This material has the advantage of a low neutron capture cross-section, but has two major disadvantages:

- It limits the maximum temperature, and hence the thermal efficiency, of the plant.
- It reacts with water, preventing long-term storage of spent fuel under water.

Magnox fuel incorporated cooling fins to provide maximum heat transfer despite low operating temperatures, making it expensive to produce. While the use of uranium metal rather than oxide made reprocessing more straightforward and therefore cheaper, the need to reprocess fuel a short time after removal from the reactor meant that the fission product hazard was severe. Expensive remote handling facilities were required to address this danger.

TRISO Fuel

Tristructural-isotropic (TRISO) fuel is a type of micro fuel particle. It consists of a fuel kernel composed of UO_X (sometimes UC or UCO) in the center, coated with four layers of three isotropic materials. The four layers are a porous buffer layer made of carbon, followed by a dense inner layer of pyrolytic carbon (PyC), followed by a ceramic layer of SiC to retain fission products at elevated temperatures and to give the TRISO particle more structural integrity, followed by a dense outer layer of PyC. TRISO fuel particles are designed not to crack due to the stresses from processes (such as differential thermal expansion or fission gas pressure) at temperatures beyond 1600°C, and therefore can contain the fuel in the worst of accident scenarios in a properly designed reactor. Two such reactor designs are the pebble bed reactor (PBR), in which thousands of TRISO fuel particles are dispersed into graphite pebbles, and the prismatic-block gas-cooled reactor (such as the GT-MHR), in which the TRISO fuel particles are fabricated into compacts and placed in a graphite block matrix. Both of these reactor designs are very high temperature reactors (VHTR) [formally known as the high-temperature gas-cooled reactors (HTGR)], one of the six classes of reactor designs in the Generation IV initiative. TRISO fuel particles were originally developed in Germany for high-temperature gas-cooled reactors. The first nuclear reactor to use TRISO fuels was the AVR and the first powerplant was the THTR-300. Currently, TRISO fuel compacts are being used in the experimental reactors, the HTR-10 in China, and the HTTR in Japan.

QUADRISO Fuel

In QUADRISO particles a burnable neutron poison (europium oxide or erbium oxide or carbide) layer surrounds the fuel kernel of ordinary TRISO particles to better manage the excess of reactivity. If the core is equipped both with TRISO and QUADRISO fuels, at beginning of life neutrons do not reach the fuel of the QUADRISO particles because they are stopped by the burnable poison. After irradiation the poison depletes and neutrons streams into the fuel kernel of QUADRISO particles inducing fission reactions. This mechanism compensates fuel depletion of ordinary TRISO fuel. In the generalized QUADRISO fuel concept the poison can eventually be mixed with the fuel kernel or the outer pyrocarbon. The QUADRISO concept has been conceived at Argonne National Laboratory.

RBMK Fuel

RBMK reactor fuel was used in Soviet designed and built RBMK type reactors. This is a low enriched uranium oxide fuel. The fuel elements in an RBMK are 3 m long each, and two of these sit back-to-back on each fuel channel, pressure tube. Reprocessed uranium from Russian VVER reactor spent fuel is used to fabricate RBMK fuel. Following the Chernobyl accident, the enrichment of fuel was changed from 2.0% to 2.4%, to compensate for control rod modifications and the introduction of additional absorbers.

CerMet Fuel

CerMet fuel consists of ceramic fuel particles (usually uranium oxide) embedded in a metal matrix. It is hypothesized that this type of fuel is what is used in United States Navy reactors. This fuel has high heat transport characteristics and can withstand a large amount of expansion.

Plate Type Fuel

Plate type fuel has fallen out of favour over the years. Plate type fuel is commonly composed of enriched uranium sandwiched between metal cladding. Plate type fuel is used in several research reactors where a high neutron flux is desired, for uses such as material irradiation studies or isotope production, without the high temperatures seen in ceramic, cylindrical fuel. It is currently used in the Advanced Test Reactor (ATR) at Idaho National Laboratory.

Sodium Bonded Fuel

Sodium bonded fuel consists of fuel that has liquid sodium in the gap between the fuel slug (or pellet) and the cladding. This fuel type is often used for sodium cooled liquid metal fast reactors. It has been used in EBR-I, EBR-II, and the FFTF. The fuel slug may be metallic or ceramic. The sodium bonding is used to reduce the temperature of the fuel.

Spent Nuclear Fuel

Used nuclear fuel is a complex mixture of the fission products, uranium, plutonium and the transplutonium metals. In fuel which has been used at high temperature in power reactors it is common for the fuel to be *heterogeneous*; often the fuel will contain nanoparticles of platinum group metals such as palladium. Also the fuel may well have cracked, swelled and been used close to its melting point. Despite the fact that the used fuel can be cracked, it is very insoluble in water, and is able to retain the vast majority of the actinides and fission products within the uranium dioxide crystal lattice.

Oxide Fuel Under Accident Conditions

Two main modes of release exist, the fission products can be vaporised or small particles of the fuel can be dispersed.

Fuel Behaviour and Post Irradiation Examination (PIE)

Materials in a high radiation environment (such as a reactor) can undergo unique behaviours such as swelling and non-thermal creep. If there are nuclear reactions within the material (such as what happens in the fuel), the stoichiometry will also change slowly over time. These behaviours can lead to new material properties, cracking, and fission gas release:

- Fission gas release
  - As the fuel is degraded or heated the more volatile fission products which are trapped within the uranium dioxide may become free. For example see J.Y. Colle, J.P. Hiernaut, D. Papaioannou, C. Ronchi, A. Sasahara, *Journal of Nuclear Materials*, 2006, 348, 229.
- Fuel cracking
  - As the fuel expands on heating, the core of the pellet expands more than the rim which may lead to cracking. Because of the thermal stress thus formed the fuel cracks, the cracks tend to go from the center to the edge in a star-shaped pattern.

In order to better understand and control these changes in materials, these behaviours are studied. A common experiment to do this is post irradiation examination, in which fuel will be examined after it is put through reactor-like conditions. Due to the intensely radioactive nature of the used fuel this is done in a hot cell. A combination of nondestructive and destructive methods of PIE are common. The PIE is used to check that the fuel is both safe and effective. After major accidents, the core is normally subject to PIE in order to find out what happened. One site where PIE is done is the ITU which is the EU center for the study of highly radioactive materials. In addition to the effects of radiation and the fission products on materials, scientists also need to consider the temperature of materials in a reactor, and in particular, the fuel. Too high a fuel temperature can compromise the fuel, and therefore it is important to control the temperature in order to control the fission chain reaction. The temperature of the fuel varies as a function of the distance from the center to the rim. At distance x from the center the temperature (T_x) is described by the equation where ρ is the power density (W m^{-3}) and K_f is the thermal conductivity.

$$T_x = T_{Rim} + \rho\,(r_{pellet}^{\;2} - x^2)\,(4\,K_f)^{-1}$$

To explain this for a series of fuel pellets being used with a rim temperature of 200 °C (typical for a BWR) with different diameters and power densities of 250 MW m^{-3} have been modeled using the above equation. Note that these fuel pellets are rather large; it is normal to use oxide pellets which are about 10 mm in diameter.

Radioisotope Decay Fuels

Radioisotope Battery: The terms atomic battery, nuclear battery and radioisotope battery are used interchangely to describe a device which uses the radioactive decay to generate electricity. These systems use radioisotopes that produce low energy beta particles or sometimes alpha particles of varying energies. Low energy beta particles are needed to prevent the production of high energy penetrating Bremsstrahlung radiation that would require heavy shielding. Radioisotopes such as tritium, nickel-63, promethium-147, and technetium-99 have been tested. Plutonium-238, curium-242, curium-244 and strontium-90 have been used. There are two main categories of atomic batteries: thermal and non-thermal. The non-thermal atomic batteries, which have many different designs, exploit charged alpha and beta particles. These designs include the direct charging generators, betavoltaics, the optoelectric nuclear battery, and the radioisotope piezoelectric generator. The thermal atomic batteries on the other hand, convert the heat from the radioactive decay to electricity. These designs include thermionic converter, thermophotovoltaic cells, alkali-metal thermal to electric converter, and the most common design, the radioisotope thermoelectric generator.

Radioisotope Heater Units (RHU)

Radioisotope heater units normally provide about 1 watt of heat each, derived from the decay of a few grams of plutonium-238. This heat is given off continuously for several decades. Their function is to provide highly localised heating of sensitive equipment (such as electronics in outer space). The Cassini-Huygens orbiter to Saturn contains 82 of these units (in addition to its 3 main RTG's for power generation). The Huygens probe to Titan contains 35 devices.

Fusion Fuels

Fusion fuels include tritium (^3H) and deuterium (^2H) as well as helium-3 (^3He). Many other elements can be fused together, but the larger electrical charge of their nuclei means that much higher temperatures are required. Only the fusion of the lightest elements is seriously considered as a future energy source. Although the energy density of fusion fuel is even higher than fission fuel, and fusion reactions sustained for a few minutes have been achieved, utilizing fusion fuel as a net energy source remains a theoretical possibility.

First Generation Fusion Fuel

Deuterium and tritium are both considered first-generation fusion fuels; they are the easiest to fuse, because the electrical charge on their nuclei is the lowest of all elements. The three most commonly cited nuclear reactions that could be used to generate energy are:

$$^2H + {}^3H\ n\ (14.07\ MeV) + {}^4He\ (3.52\ MeV)$$

$$^2H + {}^2H\ n\ (2.45\ MeV) + {}^3He\ (0.82\ MeV)$$

$$^2H + {}^2H\ p\ (3.02\ MeV) + {}^3H\ (1.01\ MeV)$$

Second Generation Fusion Fuel

Second generation fuels require either higher confinement temperatures or longer confinement time than those required of first generation fusion fuels, but generate fewer neutrons. Neutrons are an unwanted byproduct of fusion reactions in an energy generation context, because they are absorbed by the walls of a fusion chamber, making them radioactive. They cannot be confined by magnetic fields, because they are not electrically charged. This group consists of deuterium and helium-3. The products are all charged particles, but there may be significant side reactions leading to the production of neutrons.

$$^2H + {}^3He \longrightarrow p\ (14.68\ MeV) + {}^4He\ (3.67\ MeV)$$

Third Generation Fusion Fuel

Third generation fusion fuels produce only charged particles in the primary reactions, and side reactions are relatively unimportant. Since a very small amount of neutrons is produced, there would be little induced radioactivity in the walls of the fusion chamber. This is often seen as the end goal of fusion research. ^3He has the highest Maxwellian reactivity of any 3rd generation fusion fuel. However, there are no significant natural sources of this substance on Earth.

$$^3He + {}^3He \longrightarrow 2p + {}^4He\ (12.86\ MeV)$$

Another potential aneutronic fusion reaction is the proton-boron reaction:

$$p + {}^{11}B \longrightarrow 3\,{}^{4}He$$

Under reasonable assumptions, side reactions will result in about 0.1% of the fusion power being carried by neutrons. With 123 keV, the optimum temperature for this reaction is nearly ten times higher than that for the pure hydrogen reactions, the energy confinement must be 500 times better than that required for the D-T reaction, and the power density will be 2500 times lower than for D-T.

Binding Energy

Binding energy is the mechanical energy required to disassemble a whole into separate parts. A bound system typically has a lower potential energy than its constituent parts; this is what keeps the system together—often this means that energy is released upon the creation of a bound state. The usual convention is that this corresponds to a *positive* binding energy. In general, binding energy represents the mechanical work which must be done against the forces which hold an object together, disassembling the object into component parts separated by sufficient distance that further separation requires negligible additional work.

At the atomic level the atomic binding energy of the atom derives from electromagnetic interaction and is the energy required to disassemble an atom into free electrons and a nucleus. Electron binding energy is a measure of the energy required to free electrons from their atomic orbits. This is more commonly known as ionization energy. At the nuclear level, binding energy is also equivalent to the energy liberated when a nucleus is created from other nucleons or nuclei. This nuclear binding energy (binding energy of nucleons into a nuclide) is derived from the strong nuclear force and is the energy required to disassemble a nucleus into the same number of free unbound neutrons and protons it is composed of, so that the nucleons are far/distant enough from each other so that the strong nuclear force can no longer cause the particles to interact. In astrophysics, gravitational binding energy of a celestial body is the energy required to expand the material to infinity. This quantity is not to be confused with the gravitational potential energy, which is the energy required to separate two bodies, such as a celestial body and a satellite, to infinite distance, keeping each intact (the latter energy is lower). In bound systems, if the binding energy is removed from the system, it must be subtracted from the mass of the unbound system, simply because this energy has mass, and if subtracted from the system at the time it is bound, will result in removal of mass from the system. System mass is not conserved in this process because the system is not closed during the binding process.

Mass Deficit

Classically a bound system is at a lower energy level than its unbound constituents, its mass must be less than the total mass of its unbound constituents. For systems with low binding energies, this "lost" mass after binding may be fractionally small. For systems with high binding energies, however, the missing mass may be an easily measurable fraction. Since all forms of energy have mass, the question of where the missing mass of the binding energy goes, is of interest.

The answer is that this mass is lost from a system which is not closed. It transforms to heat, light, higher energy states of the nucleus/atom or other forms of energy, but these types of energy also have mass, and it is necessary that they be removed from the system before its mass may decrease. The "mass deficit" from binding energy is therefore removed mass that corresponds with removed energy, according to Einstein's equation $E = mc^2$. Once the system cools to normal temperatures and returns to ground states in terms of energy levels, there is less mass remaining in the system than there was when it first combined and was at high energy. Mass measurements are almost always made at low temperatures with systems in ground states, and this difference between the mass of a system and the sum of the masses of its isolated parts is called a mass deficit. Thus, if binding energy mass is transformed into heat, the system must be cooled (the heat removed) before the mass-deficit appears in the cooled system. In that case, the removed heat represents exactly the mass "deficit", and the heat itself retains the mass which was lost.

As an illustration, consider two objects attracting each other in space through their gravitational field. The attraction force accelerates the objects and they gain some speed toward each other converting the potential (gravity) energy into kinetic (movement) energy. When either the particles 1) pass through each other without interaction or 2) elastically repel during the collision, the gained kinetic energy (related to speed), starts to revert into potential form driving the collided particles apart.

The decelerating particles will return to the initial distance and beyond into infinity or stop and repeat the collision (oscillation takes place). This shows that the system, which loses no energy, does not combine (bind) into a solid object, parts of which oscillate at short distances. Therefore, in order to bind the particles, the kinetic energy gained due to the attraction must be dissipated (by resistive force). Complex objects in collision ordinarily undergo inelastic collision, transforming some kinetic energy into internal energy (heat content, which is atomic movement), which is further radiated in the form of photons—the light and heat. Once the energy to escape the gravity is dissipated in the collision, the parts will oscillate at closer, possibly atomic, distance, thus looking like one solid object. This lost energy, necessary to overcome the potential barrier in order to separate the objects, is the binding energy. If this binding energy were retained in the system as heat, its mass would not decrease. However, binding energy lost from the system (as heat radiation) would itself have mass, and directly represent of the "mass deficit" of the cold, bound system.

Closely analogous considerations apply in chemical and nuclear considerations. Exothermic chemical reactions in closed systems do not change mass, but (in theory) become less massive once the heat of reaction is removed. This mass change is too small to measure with standard equipment. In nuclear reactions, however, the fraction of mass that may be removed as light or heat, i.e., binding energy, is often a much larger fraction of the system mass. It may thus be measured directly as a mass difference between rest masses of reactants and products. This is because nuclear forces are comparatively stronger than Coulombic forces associated with the interactions between electrons and protons that generate heat in chemistry.

Mass Defect

In simple words, the definition of mass defect can be stated as follows:

The difference between the unbound system calculated mass and experimentally measured mass of nucleus is called mass defect. It is denoted by Äm. It can be calculated as follows:

Mass defect = (unbound system calculated mass) - (measured mass of nucleus)

i.e, (sum of masses of protons and neutrons) - (measured mass of nucleus)

In nuclear reactions, the energy that must be radiated or otherwise removed as binding energy may be in the form of electromagnetic waves, such as gamma radiation, or as heat. Again, however, no mass deficit can in theory appear until this radiation has been emitted and is no longer part of the system. The energy given off during either nuclear fusion or nuclear fission is the difference between the binding energies of the fuel and the fusion or fission products. In practice, this energy may also be calculated from the substantial mass differences between the fuel and products, once evolved heat and radiation have been removed. When the nucleons are grouped together to form a nucleus, they lose a small amount of mass, i.e., there is mass defect. This mass defect is released as (often radiant) energy according to the relation $E = mc^2$; thus binding energy = mass defect × c^2. This energy is a measure of the forces that hold the nucleons together, and it represents energy which must be supplied from the environment if the nucleus is to be broken up. It is known as binding energy, and the mass defect is a measure of the binding energy because it simply represents the mass of the energy which has been lost to the environment after binding.

Mass Excess

It is observed experimentally that the mass of the nucleus is smaller than the number of nucleons each counted with a mass of 1 a.m.u.. This difference is called mass excess. The difference between the actual mass of the nucleus measured in atomic mass units and the number of nucleons is called mass excess i.e.

Mass excess $= M - A =$ *Excess-energy* / c^2

with : M equals the actual mass of the nucleus, in u.

and : A equals the mass number.

This mass excess is a practical value calculated from experimentally measured nucleon masses and stored in nuclear databases. For middle-weight nuclides this value is negative in contrast to the mass defect which is never negative for any nuclide.

Chapter 6

Nuclear Forces

Nuclear Reaction

In nuclear physics and nuclear chemistry, a nuclear reaction is the process in which two nuclei, or else a nucleus of an atom and a subatomic particle (such as a proton, or high energy electron) from outside the atom, collide to produce products different from the initial particles. In principle, a reaction can involve more than three particles colliding, but because the probability of three or more nuclei to meet at the same time at the same place is much less than for two nuclei, such an event is exceptionally rare. While the transformation is spontaneous in the case of radioactive decay, it is initiated by a particle in the case of a nuclear reaction. If the particles collide and separate without changing, the process is called an elastic collision rather than a reaction.

Natural nuclear reactions occur in the interaction between cosmic rays and matter. On Earth, nuclear reactions initiated by radioactive decay produce nucleogenic isotopes rather than the radiogenic daughter isotopes of simple decay. Unlike radioactive decay, nuclear reactions can be employed artificially to obtain nuclear energy, at an adjustable rate, on demand. Perhaps the most notable nuclear reactions are the nuclear chain reaction that produces nuclear fission, and the nuclear fusion reactions that power the energy production of the Sun and stars. Both of these types of reactions are employed in nuclear weapons.

Notation

Instead of using the full equations as shown in the previous section, in many situations a compact notation is used to describe nuclear reactions. This is A(b,c)D, which is equivalent to A + b gives c + D. Common light particles are often abbreviated in this shorthand, typically p for proton, n for neutron, d for deuteron, α representing an alpha particle or helium-4, β for beta particle or electron, γ for gamma photon, etc. The reaction above would be written as Li-6(d,α)α.

Energy Conservation

Kinetic energy may be released during the course of a reaction (exothermic reaction) or kinetic energy may have to be supplied for the reaction to take place (endothermic reaction). This can be calculated by reference to a table of very accurate particle rest masses <http://physics.nist.gov/PhysRefData/Compositions/index.html>, as follows.

- Total rest mass on left side = 6.015 + 2.014 = 8.029 u
- Total rest mass on right side = 2 × 4.0026 = 8.0052 u
- Missing rest mass = 8.029 – 8.0052 = 0.0238 atomic mass units.

In a nuclear reaction, the total (relativistic) energy is conserved. The "missing" rest mass must therefore reappear as kinetic energy released in the reaction; its source is the nuclear binding energy. Using Einstein's mass-energy equivalence formula $E = mc^2$, the amount of energy released can be determined. We first need the energy equivalent of one atomic mass unit:

$$1 \text{ u } c^2 = (1.66054 \times 10^{-27} \text{ kg}) \times (2.99792 \times 10^8 \text{ m/s})^2$$

$$= 1.49242 \times 10^{-10} \text{ kg (m/s)}^2 = 1.49242 \times 10^{"10} \text{ J (Joule)}$$

$$\times (1 \text{ MeV} / 1.60218 \times 10^{"13} \text{ J})$$

$$= 931.49 \text{ MeV},$$

$$\text{so } 1 \text{ u } c^2 = 931.49 \text{ MeV}.$$

Hence, the energy released is 0.0238 × 931 MeV = 22.4 MeV. Expressed differently: the mass is reduced by 0.3 %, corresponding to 0.3 % of 90 PJ/kg is 300 TJ/kg. This is a large amount of energy for a nuclear reaction; the amount is so high because the binding energy per nucleon of the helium-4 nucleus is unusually high, because the He-4 nucleus is "doubly magic". (The He-4 nucleus is unusually stable and tightly-bound for the same reason that the helium atom is inert: each pair of protons and neutrons in He-4 occupies a filled 1s nuclear orbital in the same way that the pair of electrons in the helium atom occupy a filled 1s electron orbital). Consequently, alpha particles appear frequently on the right hand side of nuclear reactions. The energy released in a nuclear reaction can appear mainly in one of three ways:

- kinetic energy of the product particles
- emission of very high energy photons, called gamma rays
- some energy may remain in the nucleus, as a metastable energy level.

When the product nucleus is metastable, this is indicated by placing an asterisk ("*") next to its atomic number. This energy is eventually released through nuclear decay. A small amount of energy may also emerge in the form of X-rays. Generally, the product nucleus has a different atomic number, and thus the configuration of its electron shells is wrong. As the electrons rearrange themselves and drop to lower energy levels, internal transition X-rays (X-rays with precisely defined emission lines) may be emitted.

Q-value and Energy Balance

In writing down the reaction equation, in a way analogous to a chemical equation, one may in addition give the reaction energy on the right side:

Target nucleus + projectile —> Final nucleus + ejectile + Q.

For the particular case discussed above, the reaction energy has already been calculated as Q = 22.4 MeV. The reaction energy (the "Q-value") is positive for exothermal reactions and negative for endothermal reactions. On the one hand, it is the difference between the sums of kinetic energies on the final side and on the initial side. But on the other hand, it is also the difference between the nuclear rest masses on the initial side and on the final side (in this way, we have calculated the Q-value above).

Reaction Rates

If the reaction equation is balanced, that does not mean that the reaction really occurs. The rate at which reactions occur depends on the particle energy, the particle flux and the reaction cross section. An example of a large repository of reaction rates is the REACLIB database, as maintained by the Joint Institute for Nuclear Astrophysics.

Neutrons vs Ions

In the initial collision which begins the reaction, the particles must approach closely enough so that the short range strong force can affect them. As most common nuclear particles are positively charged, this means they must overcome considerable electrostatic repulsion before the reaction can begin. Even if the target nucleus is part of a neutral atom, the other particle must penetrate well beyond the electron cloud and closely approach the nucleus, which is positively charged. Thus, such particles must be first accelerated to high energy, for example by:

- particle accelerators
- nuclear decay (alpha particles are the main type of interest here, since beta and gamma rays are rarely involved in nuclear reactions)
- very high temperatures, on the order of millions of degrees, producing thermonuclear reactions
- cosmic rays

Also, since the force of repulsion is proportional to the product of the two charges, reactions between heavy nuclei are rarer, and require higher initiating energy, than those between a heavy and light nucleus; while reactions between two light nuclei are the most common ones. Neutrons, on the other hand, have no electric charge to cause repulsion, and are able to effect a nuclear reaction at very low energies.

In fact, at extremely low particle energies (corresponding, say, to thermal equilibrium at room temperature), the neutron's de Broglie wavelength is greatly increased, possibly greatly increasing its capture cross section, at energies close to resonances of the nuclei involved. Thus low energy neutrons *may* be even more reactive than high energy neutrons.

Notable Types

While the number of possible nuclear reactions is immense, there are several types which are more common, or otherwise notable. Some examples include:

- Fusion reactions — two light nuclei join to form a heavier one, with additional particles (usually protons or neutrons) thrown off to conserve momentum.
- Fission reactions — a very heavy nucleus, spontaneously or after absorbing additional light particles (usually neutrons), splits into two or sometimes three pieces. (a decay is not usually called fission.)
- Spallation — a nucleus is hit by a particle with sufficient energy and momentum to knock out several small fragments or, smash it into many fragments.
- Induced gamma emission belongs to a class in which only photons were involved in creating and destroying states of nuclear excitation.

Direct Reactions

An intermediate energy projectile transfers energy or picks up or loses nucleons to the nucleus in a single quick (10^{-21} second) event. Energy and momentum transfer are relatively small.

These are particularly useful in experimental nuclear physics, because the reaction mechanisms are often simple enough to calculate with sufficient accuracy to probe the structure of the target nucleus.

Inelastic Scattering

Only energy and momentum are transferred.

- (p,p') tests differences between nuclear states
- (a,a') measures nuclear surface shapes and sizes. Since a particles that hit the nucleus react more violently, elastic and shallow inelastic a scattering are sensitive to the shapes and sizes of the targets, like light scattered from a small black object.
- (e,e') is useful for probing the interior structure. Since electrons interact less strongly than do protons and neutrons, they reach to the centers of the targets and their wave functions are less distorted by passing through the nucleus.

Transfer Reactions

Usually at moderately low energy, one or more nucleons are transferred between the projectile and target. These are useful in studying outer shell structure of nuclei.

- (α,n) and (a,p) reactions. Some of the earliest nuclear reactions studied involved an alpha particle produced by alpha decay, knocking a nucleon from a target nucleus.
- (d,n) and (d,p) reactions. A deuteron beam impinges on a target; the target nuclei absorb either the neutron or proton from the deuteron. The deuteron is so loosely bound that this is almost the same as proton or neutron capture. A compound nucleus may be formed, leading to additional neutrons being emitted more slowly. (d,n) reactions are used to generate energetic neutrons.
- The strangeness exchange reaction (K, π) has been used to study hypernuclei.
- The reaction $^{14}N(a,p)^{17}O$ performed by Rutherford in 1919, is generally regarded as the first nuclear transmutation experiment.

Some reactions are only possible with fast neutrons:

- (n,2n) reactions produce small amounts of protactinium-231 and uranium-232 in the thorium cycle which is otherwise relatively free of highly radioactive actinide products.
- $^{9}Be + n \longrightarrow 2a + 2n$ can contribute some additional neutrons in the beryllium neutron reflector of a nuclear weapon.
- $^{7}Li + n \longrightarrow T + a + n$ unexpectedly contributed additional yield in Castle Bravo, Castle Romeo, and Castle Yankee, the three highest-yield nuclear tests conducted by the U.S.

Compound Nuclear Reactions

Either a low energy projectile is absorbed or a higher energy particle transfers energy to the nucleus, leaving it with too much energy to be fully bound together. On a time scale of about 10^{-19} seconds, particles, usually neutrons, are "boiled" off. That is, it remains together until enough energy happens to be concentrated in one neutron to escape the mutual attraction. Charged particles rarely boil off because of the coulomb barrier. The excited quasi-bound nucleus is called a compound nucleus.

- Low energy (e, e' xn), (γ, xn) (the xn indicating one or more neutrons), where the gamma or virtual gamma energy is near the giant dipole resonance. These increase the need for radiation shielding around electron accelerators

Nuclear Force

The nuclear force (or nucleon-nucleon interaction or residual strong force) is the force between two or more nucleons. It is responsible for binding of protons and neutrons into atomic nuclei. The energy released causes the masses of nuclei to be less than the total mass of the protons and neutrons which form them.

The nuclear force is now understood as a residual effect of the even more powerful strong force, or strong interaction, which is the attractive force that binds particles called quarks together, to form the nucleons themselves. This more powerful force is mediated by particles called gluons. Gluons hold quarks together with a force like that of electric charge, but of far greater power. The concept of a nuclear force was first quantitatively constructed in 1934, shortly after the discovery of the neutron revealed that atomic nuclei were made of protons and neutrons, held together by an attractive force. The nuclear force at that time was conceived to be transmitted by particles called mesons, which were predicted in theory before being discovered in 1947.

In the 1970's, further understanding revealed these mesons to be combinations of quarks and gluons, transmitted between nucleons that themselves were made of quarks and gluons. This new model allowed the strong forces that held nucleons together, to be felt in neighboring nucleons, as residual strong forces.

The nuclear forces arising between nucleons are now seen to be analogous to the forces in chemistry between neutral atoms called van der Waals forces. Such forces between atoms are much weaker than the electrical forces that hold the atoms themselves together, and their range is shorter, because they arise from spontaneous separation of charges inside the atom.

Similarly, even though nucleons are made of quarks and gluons that are in combinations which cancel most gluon forces, some combinations of quarks and gluons nevertheless leak away from nucleons, in the form of short-range nuclear force fields that extend from one nucleon to another close by. These nuclear forces are very weak compared to direct gluon forces inside nucleons, and they extend only over a few nuclear diameters, falling exponentially with distance. Nevertheless, they are strong enough to bind neutrons and protons over short distances, and overcome the electrical repulsion between protons in the nucleus.

History

The nuclear force has been at the heart of nuclear physics ever since the field was born in 1932 with the discovery of the neutron by James Chadwick. The traditional goal of nuclear physics is to understand the properties of atomic nuclei in terms of the 'bare' interaction between pairs of nucleons, or nucleon-nucleon forces (*NN* forces). In 1934, Hideki Yukawa made the earliest attempt to explain the nature of the nuclear force.

According to his theory, massive bosons (mesons) mediate the interaction between two nucleons. Although, in light of QCD, meson theory is no longer perceived as fundamental, the meson-exchange concept (where hadrons are treated as elementary particles) continues to represent the best working model for a quantitative *NN* potential. Historically, it was a formidable task to describe the nuclear force phenomenologically, and the first semi-empirical quantitative models came in the mid-1950s. There has been substantial progress in experiment and theory related to the nuclear force. Most basic questions were settled in the 1960s and 1970s. In recent years, experimenters have concentrated on the subtleties of the nuclear force, such as its charge dependence, the precise value of the πNN coupling constant, improved phase shift analysis, high-precision NN data, high-precision NN potentials, NN scattering at intermediate and high energies, and attempts to derive the nuclear force from QCD.

To a large extent, the nuclear force can be understood in terms of the exchange of virtual light mesons, such as the pions. Sometimes the nuclear force is called the residual strong force, in contrast to the strong interactions which are now understood to arise from quantum chromodynamics (QCD).

This phrasing arose during the 1970s when QCD was being established. Before that time, the *strong nuclear force* referred to the inter-nucleon potential. After the verification of the quark model, *strong interaction* has come to mean QCD.

Basic Properties of the Nuclear Force

The nuclear force is only felt among hadrons. At much smaller separations between nucleons the force is very powerfully repulsive, which keeps the nucleons at a certain average separation. Beyond about 1.7 femtometer (fm) separation, the force drops to negligibly small values. At short distances, the nuclear force is stronger than the Coulomb force; it can overcome the Coulomb repulsion of protons inside the nucleus.

However, the Coulomb force between protons has a much larger range and becomes the only significant force between protons when their separation exceeds about 2.5 fm. To disassemble a nucleus into unbound protons and neutrons would require doing work against the nuclear force. Conversely, energy is released when a nucleus is created from other nucleons or nuclei: the nuclear binding energy. Because of mass–energy equivalence (i.e. Einstein's famous formula $E = mc^2$), releasing this energy causes the mass of the nucleus to be lower than the total mass of the individual nucleons, leading to the so-called "mass defect". The nuclear force is nearly independent of whether the nucleons are neutrons or protons. This property is called *charge independence.* It depends on whether the spins of the nucleons are parallel or antiparallel, and has a noncentral or *tensor* component. This part of the force does not conserve orbital angular momentum, which is a constant of motion under central forces.

Since nucleons have no color charge, the nuclear force does not *directly* involve the force carriers of quantum chromodynamics, the gluons. However, just as electrically neutral atoms (each composed of cancelling charges) attract each other via the second-order effects of electrical polarization, via the van der Waals forces (London forces), so by analogy, "color-neutral" nucleons may attract each other by a type of polarization which allows some basically gluon-mediated effects to be carried from one color-neutral nucleon to another, via the virtual mesons which transmit the forces, and which themselves are held together by virtual gluons.

It is this *van der Waals-like* nature which is responsible for the term "residual" in the term "residual strong force." The basic idea is that while the nucleons are "color-neutral," just as atoms are "charge-neutral," in both cases, polarization effects acting between near-by neutral particles allow a "residual" charge effect to cause net charge-mediated attraction between uncharged species, although it is necessarily of a much weaker and less direct nature than the basic forces which act internally within the particles.

Nucleon-Nucleon Potentials

Two-nucleon systems such as the deuteron, the nucleus of a deuterium atom, as well as proton-proton or neutron-proton scattering are ideal for studying the *NN* force. Such systems can be described by attributing a *potential* (such as the Yukawa potential) to the nucleons and using the potentials in a Schrödinger equation.

The form of the potential is derived phenomenologically, although for the long-range interaction, meson-exchange theories help to construct the potential. The parameters of the potential are determined by fitting to experimental data such as the deuteron binding energy or *NN* elastic scattering cross sections (or, equivalently in this context, so-called *NN* phase shifts).

The most widely used *NN* potentials are the Paris potential, the Argonne AV18 potential, the CD-Bonn potential and the Nijmegen potentials. A more recent approach is to develop effective field theories for a consistent description of nucleon-nucleon and three-nucleon forces.

In particular, chiral symmetry breaking can be analysed in terms of an effective field theory (called chiral perturbation theory) which allows perturbative calculations of the interactions between nucleons with pions as exchange particles.

From Nucleons to Nuclei

The ultimate goal of nuclear physics would be to describe all nuclear interactions from the basic interactions between nucleons. This is called the *microscopic* or *ab initio* approach of nuclear physics. There are two major obstacles to overcome before this dream can become reality:

- Calculations in many-body systems are difficult and require advanced computation techniques.
- There is evidence that three-nucleon forces (and possibly higher multi-particle interactions) play a significant role. This means that three-nucleon potentials must be included into the model.

This is an active area of research with ongoing advances in computational techniques leading to better first-principles calculations of the nuclear shell structure. Two- and three-nucleon potentials have been implemented for nuclear masses up to A=12.

Nuclear Potentials

A successful way of describing nuclear interactions is to construct one potential for the whole nucleus instead of considering all its nucleon components. This is called the *macroscopic* approach. For example, scattering of neutrons from nuclei can be described by considering a plane wave in the potential of the nucleus, which comprises a real part and an imaginary part. This model is often called the optical model since it resembles the case of light scattered by an opaque glass sphere. Nuclear potentials can be *local* or *global*: local potentials are limited to a narrow energy range and/or a narrow nuclear mass range, while global potentials, which have more parameters and are usually less accurate, are functions of the energy and the nuclear mass and can therefore be used in a wider range of applications.

Strong Interaction

In particle physics, the strong interaction (also called the strong force, strong nuclear force, or colour force) is one of the four fundamental interactions of nature, the others being electromagnetism, the weak interaction and gravitation. It is a non-contact force. At atomic scale, it is about 100 times stronger than electromagnetism, which in turn is orders of magnitude stronger than the weak force and gravitation. The strong interaction is observable in two areas: On the larger scale, it is the force that binds protons and neutrons together to form the nucleus of an atom. On the smaller scale, it is also the force that holds quarks and gluons together to form the proton, the neutron and other particles.

In the context of binding protons and neutrons (nucleons) together to form atoms, the strong interaction is called the nuclear force (or *residual strong force*). In this case, it is the residuum of the strong interaction between the quarks that make up the protons and neutrons. As such, the residual strong interaction obeys a quite different distance-dependent behaviour between nucleons, from when it is acting to bind quarks within nucleons. The strong force is thought to be mediated by gluons, acting upon quarks, antiquarks, and the gluons themselves. This is detailed in the theory of quantum chromodynamics (QCD).

History

Before the 1970s, physicists were uncertain about the binding mechanism of the atomic nucleus. It was known that the nucleus was composed of protons and neutrons and that protons possessed positive electric charge while neutrons were electrically neutral. However, these facts seemed to contradict one another.

By physical understanding at that time, positive charges would repel one another and the nucleus should therefore fly apart. However, this was never observed. New physics was needed to explain this phenomenon. A stronger attractive force was postulated to explain how the atomic nucleus was bound together despite the protons' mutual electromagnetic repulsion. This hypothesized force was called the *strong force*, which was believed to be a fundamental force that acted on the nucleons (the protons and neutrons that make up the nucleus). Experiments suggested that this force bound protons and neutrons together with equal strength. It was later discovered that protons and neutrons were not fundamental particles, but were made up of constituent particles called quarks. The strong attraction between nucleons was the side-effect of a more fundamental force that bound the quarks together in the protons and neutrons. The theory of quantum chromodynamics explains that quarks carry what is called a color charge, although it has no relation to visible color. Quarks with unlike color charge attract one another as a result of the strong interaction, which is mediated by particles called gluons.

Details

The word *strong* is used since the strong interaction is the "strongest" of the four fundamental forces; its strength is 100 times that of the electromagnetic force, some 10^{13} times as great as that of the weak force, and about 10^{39} times that of gravitation.

The Behaviour of the Strong Force

The contemporary strong force is described by quantum chromodynamics (QCD), a part of the standard model of particle physics. Mathematically, QCD is a non-Abelian gauge theory based on a local (gauge) symmetry group called SU(3). Quarks and gluons are the only fundamental particles which carry non-vanishing color charge, and hence participate in strong interactions.

The strong force itself acts directly only upon elementary quark and gluon particles. All quarks and gluons in QCD interact with each other through the strong force. The strength of interaction is parametrized by the strong coupling constant. This strength is modified by the gauge color charge of the particle, a group theoretical property. The strong force acts between quarks.

Unlike all other forces (electromagnetic, weak, and gravitational), the strong force does not diminish in strength with increasing distance. After a limiting distance (about the size of a hadron) has been reached, it remains at a strength of about 10,000 newtons, no matter how much further the distance between the quarks. In QCD this phenomenon is called color confinement; it implies that only hadrons, not individual free quarks, can be observed.

The explanation is that the amount of work done against a force of 10,000 newtons (about the weight of a one-metric ton mass on the surface of the Earth) is enough to create particle-antiparticle pairs within a very short distance of an interaction. In simple terms, the very energy applied to pull two quarks apart will turn into new quarks that pair up again with the original ones. The failure of all experiments that have searched for free quarks is considered to be evidence for this phenomenon. The elementary quark and gluon particles affected are unobservable directly, but instead emerge as jets of newly created hadrons, whenever energy is deposited into a quark-quark bond, as when a quark in a proton is struck by a very fast quark (in an impacting proton) during a particle accelerator experiment. However, quark-gluon plasmas have been observed.

Residual Strong Force

The residual effect of the strong force is called the nuclear force. The nuclear force acts between hadrons, such as nucleons in atomic nuclei. This "residual strong force", acting indirectly, transmits gluons that form part of the virtual pi and rho mesons, which, in turn, transmit the nuclear force between nucleons. The residual strong force is thus a minor residuum of the strong force which binds quarks together into protons and neutrons.

This same force is much weaker *between* neutrons and protons, because it is mostly neutralized *within* them, in the same way that electromagnetic forces between neutral atoms (van der Waals forces) are much weaker than the electromagnetic forces that hold the atoms internally together. Unlike the strong force itself, the nuclear force, or residual strong force, *does* diminish in strength, strongly with distance.

The decrease is approximately as a negative exponential power of distance, though there is no simple expression known for this; see Yukawa potential. This fact, together with the less-rapid decrease of the disruptive electromagnetic force between protons with distance, causes the instability of larger atomic nuclei, such as all those with atomic numbers larger than 82.

Natural Nuclear Fission Reactor

A natural nuclear fission reactor is a uranium deposit where analysis of isotope ratios has shown that self-sustaining nuclear chain reactions have occurred. The existence of this phenomenon was discovered in 1972 at Oklo in Gabon, Africa, by French physicist Francis Perrin. The conditions under which a natural nuclear reactor could exist had been predicted in 1956 by Paul Kuroda. The conditions found were very similar to what was predicted. Oklo is the only known location for this in the world and consists of 16 sites at which self-sustaining nuclear fission reactions took place approximately 2 billion years ago, and ran for a few hundred thousand years, averaging 100 kW of power output during that time.

History

In May 1972 at the Pierrelatte uranium enrichment facility in France, routine mass spectrometry comparing UF_6 samples from the Oklo Mine, located in Gabon, Central Africa, showed a discrepancy in the amount of the ^{235}U isotope. Normally the concentration is 0.720%; these samples had only 0.717% – a significant difference. This discrepancy required explanation, as all uranium handling facilities must meticulously account for all fissionable isotopes to assure that none are diverted for weapons purposes.

Thus the French Commissariat à l'énergie atomique (CEA) began an investigation. A series of measurements of the relative abundances of the two most significant isotopes of the uranium mined at Oklo showed anomalous results compared to those obtained for uranium from other mines.

Further investigations into this uranium deposit discovered uranium ore with a ^{235}U to ^{238}U ratio as low as 0.440%. Subsequent examination of other isotopes showed similar anomalies, such as Nd and Ru as described in more detail below.

This loss in ^{235}U is exactly what happens in a nuclear reactor. A possible explanation therefore was that the uranium ore had operated as a natural fission reactor. Other observations led to the same conclusion, and on September 25, 1972, the CEA announced their finding that self-sustaining nuclear chain reactions had occurred on Earth about 2 billion years ago. Later, other natural nuclear fission reactors were discovered in the region.

Fission Product Isotope Signatures

Nd: Neodymium and other elements were found with isotopic compositions different from what is customarily found on Earth. For example, natural neodymium contains 27% ^{142}Nd; the Nd at Oklo contained less than 6% but contained more ^{143}Nd. Subtracting the natural isotopic Nd abundance from the Oklo-Nd, the isotopic composition matched that produced by the fissioning of ^{235}U.

Ru: Similar investigations into the isotopic ratios of ruthenium at Oklo found a much higher ^{99}Ru concentration than expected (27-30% vs. 12.7%). This anomaly could be explained by the decay of ^{99}Tc to ^{99}Ru. In the bar chart below the normal natural isotope signature of ruthenium is compared with that for fission product ruthenium which is the result of the fission of ^{235}U with thermal neutrons.

It is clear that the fission ruthenium has a different isotope signature. The level of ^{100}Ru in the fission product mixture is low because of a long lived (half life = 10^{19} years) isotope of molybdenum. On the time scale of when the reactors were in operation very little decay to ^{100}Ru will have occurred.

Mechanism of the Reactors

The natural nuclear reactor formed when a uranium-rich mineral deposit became inundated with groundwater that acted as a neutron moderator, and a nuclear chain reaction took place. The heat generated from the nuclear fission caused the groundwater to boil away, which slowed or stopped the reaction. After cooling of the mineral deposit, short-lived fission product poisons decayed, the water returned and the reaction started again.

These fission reactions were sustained for hundreds of thousands of years, until a chain reaction could no longer be supported. Fission of uranium normally produces five known isotopes of the fission-product gas xenon; all five have been found trapped in the remnants of the natural reactor, in varying concentrations. The concentrations of xenon isotopes, found trapped in mineral formations 2 billion years later, make it possible to calculate the specific time intervals of reactor operation: approximately 30 minutes of criticality followed by 2 hours and 30 minutes of cooling down to complete a 3-hour cycle.

A key factor that made the reaction possible was that, at the time the reactor went critical, the fissile isotope ^{235}U made up about 3% of the natural uranium, which is comparable to the amount used in some of today's reactors. (The remaining 97% was non-fissile ^{238}U.) Because ^{235}U has a shorter half life than ^{238}U, and thus decays more rapidly, the current abundance of ^{235}U in natural uranium is about 0.7%. A natural nuclear reactor is therefore no longer possible on Earth without heavy water.

The Oklo uranium ore deposits are the only known sites in which natural nuclear reactors existed. Other rich uranium ore bodies would also have had sufficient uranium to support nuclear reactions at that time, but the combination of uranium, water and physical conditions needed to support the chain reaction was unique to the Oklo ore bodies.

Another factor which probably contributed to the start of the Oklo natural nuclear reactor at 2 billion years, rather than earlier, was the increasing oxygen content in the Earth's atmosphere. Uranium is naturally present in the rocks of the earth, and the abundance of fissionable ^{235}U was at least 3% or higher at all times prior to reactor startup.

However, uranium is soluble in water only in the presence of oxygen. Therefore, the rising oxygen levels during the aging of earth may have allowed uranium to be dissolved and transported with groundwater to places where a high enough concentration could accumulate to form rich uranium ore bodies. Without the new aerobic environment available on earth at the time, these concentrations probably couldn't have taken place.

It is estimated that nuclear reactions in the uranium in centimeter- to meter-sized veins consumed about five tons of ^{235}U and elevated temperatures to a few hundred degrees Celsius. Remarkably, most of the non-volatile fission products and actinides have only moved centimeters in the veins during the last 2 billion years. This offers a case study of how radioactive isotopes migrate through the Earth's crust.

Relation to the Atomic Fine-Structure Constant

The natural reactor of Oklo has been used to check if the atomic fine-structure constant α might have changed over the past 2 billion years. That is because α influences the rate of various nuclear reactions. For example, ^{149}Sm captures a neutron to become ^{150}Sm, and since the rate of neutron capture depends on the value of α, the ratio of the two samarium isotopes in samples from Oklo can be used to calculate the value of α from 2 billion years ago.

Several studies have analysed the relative concentrations of radioactive isotopes left behind at Oklo, and most (but not all) have concluded that nuclear reactions then were much the same as they are today, which implies α was the same too. The ^{149}Sm resonance is sensitive to changes in the proton-to-electron mass ratio, ì, as well as changes in α. It may be that some cancellation occurs between these two, meaning that the above results could also have been obtained if α and ì had both been different, in the correct proportions, at the time the Oklo reactor was running.

In Fiction

In the science fiction novel *Bridge of Ashes*, by Roger Zelazny, the Gabon mine was stated to have been created by an alien race for the purpose of causing mutations which ultimately led to the existence of humans. In the novel *Camelot 30K* by Robert L. Forward, natural nuclear fission (as opposed to solar power) is revealed as the power source of the biosphere of a trans-Neptunian dwarf planet. A natural nuclear fission reactor is also an important plot device in the science fiction novel *Manifold: Space* by Stephen Baxter. In 1956, a natural nuclear fission reactor figured in *Tom Swift in the Caves of Nuclear Fire* by Victor Appleton II (house pseudonym for Thomas Mulvey).

Big Bang

The Big Bang model or theory is the prevailing cosmological theory of the early development of the universe. According to the Big Bang model, the universe was originally in an extremely hot and dense state that expanded rapidly.

This expansion caused the universe to cool and resulted in the present diluted state that continues to expand today. Based on the best available measurements as of 2010, the original state of the universe existed around 13.7 billion years ago, which is often referred to as the time when the Big Bang occurred. The theory is the most comprehensive and accurate explanation supported by scientific evidence and observations.

Monsignor Georges Lemaître, a priest from the Catholic University of Louvain, proposed what became known as the Big Bang theory of the origin of the universe, he called it his “hypothesis of the primeval atom”. The framework for the model relies on Albert Einstein’s general relativity and on simplifying assumptions (such as homogeneity and isotropy of space). The governing equations had been formulated by Alexander Friedmann.

In 1929, Edwin Hubble discovered that the distances to far away galaxies were generally proportional to their redshifts — an idea originally suggested by Lemaître in 1927. Hubble's observation was taken to indicate that all very distant galaxies and clusters have an apparent velocity directly away from our vantage point: the farther away, the higher the apparent velocity. If the distance between galaxy clusters is increasing today, everything must have been closer together in the past. This idea has been considered in detail back in time to extreme densities and temperatures, and large particle accelerators have been built to experiment on and test such conditions, resulting in significant confirmation of the theory, but these accelerators have limited capabilities to probe into such high energy regimes.

Without any evidence associated with the earliest instant of the expansion, the Big Bang theory *cannot* and *does not* provide any explanation for such an initial condition; rather, it *describes* and *explains* the general evolution of the universe since that instant. The observed abundances of the light elements throughout the cosmos closely match the calculated predictions for the formation of these elements from nuclear processes in the rapidly expanding and cooling first minutes of the universe, as logically and quantitatively detailed according to Big Bang nucleosynthesis. Fred Hoyle is credited with coining the term *Big Bang* during a 1949 radio broadcast.

It is popularly reported that Hoyle, who favoured an alternative "steady state" cosmological model, intended this to be pejorative, but Hoyle explicitly denied this and said it was just a striking image meant to highlight the difference between the two models. Hoyle later helped considerably in the effort to understand stellar nucleosynthesis, the nuclear pathway for building certain heavier elements from lighter ones. After the discovery of the cosmic microwave background radiation in 1964, and especially when its spectrum (i.e., the amount of radiation measured at each wavelength) was found to match that of thermal radiation from a black body, most scientists were fairly convinced by the evidence that some version of the Big Bang scenario must have occurred.

Motivation and Development

The Big Bang theory developed from observations of the structure of the Universe and from theoretical considerations. In 1912 Vesto Slipher measured the first Doppler shift of a "spiral nebula" (spiral nebula is the obsolete term for spiral galaxies), and soon discovered that almost all such nebulae were receding from Earth. He did not grasp the cosmological implications of this fact, and indeed at the time it was highly controversial whether or not these nebulae were "island universes" outside our Milky Way.

Ten years later, Alexander Friedmann, a Russian cosmologist and mathematician, derived the Friedmann equations from Albert Einstein's equations of general relativity, showing that the Universe might be expanding in contrast to the static Universe model advocated by Einstein at that time. In 1924, Edwin Hubble's measurement of the great distance to the nearest spiral nebulae showed that these systems were indeed other galaxies. Independently deriving Friedmann's equations in 1927, Georges Lemaître, a Belgian physicist and Roman Catholic priest, proposed that the inferred recession of the nebulae was due to the expansion of the Universe.

In 1931 Lemaître went further and suggested that the evident expansion in forward time required that the Universe contracted backwards in time, and would continue to do so until it could contract no further, bringing all the mass of the Universe into a single point, a "primeval atom" where and when the fabric of time and space comes into existence. Starting in 1924, Hubble painstakingly developed a series of distance indicators, the forerunner of the cosmic distance ladder, using the 100-inch (2,500 mm) Hooker telescope at Mount Wilson Observatory. This allowed him to estimate distances to galaxies whose redshifts had already been measured, mostly by Slipher. In 1929, Hubble discovered a correlation between distance and recession velocity—now known as Hubble's law. Lemaître had already shown that this was expected, given the Cosmological Principle.

During the 1930s other ideas were proposed as non-standard cosmologies to explain Hubble's observations, including the Milne model, the oscillatory Universe (originally suggested by Friedmann, but advocated by Albert Einstein and Richard Tolman) and Fritz Zwicky's tired light hypothesis.

After World War II, two distinct possibilities emerged. One was Fred Hoyle's steady state model, whereby new matter would be created as the Universe seemed to expand. In this model, the Universe is roughly the same at any point in time.

The other was Lemaître's Big Bang theory, advocated and developed by George Gamow, who introduced big bang nucleosynthesis (BBN) and whose associates, Ralph Alpher and Robert Herman, predicted the cosmic microwave background radiation (CMB). Ironically, it was Hoyle who coined the phrase that came to be applied to Lemaître's theory, referring to it as "this *big bang* idea" during a BBC Radio broadcast in March 1949. For a while, support was split between these two theories.

Eventually, the observational evidence, most notably from radio source counts, began to favour Big Bang over Steady State. The discovery and confirmation of the cosmic microwave background radiation in 1964 secured the Big Bang as the best theory of the origin and evolution of the cosmos.

Much of the current work in cosmology includes understanding how galaxies form in the context of the Big Bang, understanding the physics of the Universe at earlier and earlier times, and reconciling observations with the basic theory.

Huge strides in Big Bang cosmology have been made since the late 1990s as a result of major advances in telescope technology as well as the analysis of copious data from satellites such as COBE, the Hubble Space Telescope and WMAP. Cosmologists now have fairly precise and accurate measurements of many of the parameters of the Big Bang model, and have made the unexpected discovery that the expansion of the Universe appears to be accelerating.

Timeline of the Big Bang

Extrapolation of the expansion of the Universe backwards in time using general relativity yields an infinite density and temperature at a finite time in the past. This singularity signals the breakdown of general relativity. How closely we can extrapolate towards the singularity is debated—certainly not earlier than the Planck epoch. This singularity is sometimes called "the Big Bang", but the term can also refer to the early hot, dense phase is itself, which can be considered the "birth" of our Universe.

Based on measurements of the expansion using Type Ia supernovae, measurements of temperature fluctuations in the cosmic microwave background, and measurements of the correlation function of galaxies, the Universe has a calculated age of 13.75 ± 0.11 billion years. The agreement of these three independent measurements strongly supports the ËCDM model that describes in detail the contents of the Universe. The earliest phases of the Big Bang are subject to much speculation.

In the most common models, the Universe was filled homogeneously and isotropically with an incredibly high energy density, huge temperatures and pressures, and was very rapidly expanding and cooling. Approximately 10^{-37} seconds into the expansion, a phase transition caused a cosmic inflation, during which the Universe grew exponentially. After inflation stopped, the Universe consisted of a quark–gluon plasma, as well as all other elementary particles. Temperatures were so high that the random motions of particles were at relativistic speeds, and particle–antiparticle pairs of all kinds were being continuously created and destroyed in collisions. At some point an unknown reaction called baryogenesis violated the conservation of baryon number, leading to a very small excess of quarks and leptons over antiquarks and antileptons—of the order of one part in 30 million. This resulted in the predominance of matter over antimatter in the present Universe.

The Universe continued to grow in size and fall in temperature, hence the typical energy of each particle was decreasing. Symmetry breaking phase transitions put the fundamental forces of physics and the parameters of elementary particles into their present form. After about 10^{-11} seconds, the picture becomes less speculative, since particle energies drop to values that can be attained in particle physics experiments. At about 10^{-6} seconds, quarks and gluons combined to form baryons such as protons and neutrons. The small excess of quarks over antiquarks led to a small excess of baryons over antibaryons. The temperature was now no longer high enough to create new proton–antiproton pairs (similarly for neutrons–antineutrons), so a mass annihilation immediately followed, leaving just one in 10^{10} of the original protons and neutrons, and none of their antiparticles. A similar process happened at about 1 second for electrons and positrons.

After these annihilations, the remaining protons, neutrons and electrons were no longer moving relativistically and the energy density of the Universe was dominated by photons (with a minor contribution from neutrinos). A few minutes into the expansion, when the temperature was about a billion (one thousand million; 10^9; SI prefix giga-) kelvin and the density was about that of air, neutrons combined with protons to form the Universe's deuterium and helium nuclei in a process called Big Bang nucleosynthesis. Most protons remained uncombined as hydrogen nuclei. As the Universe cooled, the rest mass energy density of matter came to gravitationally dominate that of the photon radiation. After about 379,000 years the electrons and nuclei combined into atoms (mostly hydrogen); hence the radiation decoupled from matter and continued through space largely unimpeded. This relic radiation is known as the cosmic microwave background radiation.

Over a long period of time, the slightly denser regions of the nearly uniformly distributed matter gravitationally attracted nearby matter and thus grew even denser, forming gas clouds, stars, galaxies, and the other astronomical structures observable today. The details of this process depend on the amount and type of matter in the Universe. The four possible types of matter are known as cold dark matter, warm dark matter, hot dark matter and baryonic matter.

The best measurements available (from WMAP) show that the data is well-fit by a Lambda-CDM model in which dark matter is assumed to be cold (warm dark matter is ruled out by early reionization), and is estimated to make up about 23% of the matter/energy of the universe, while baryonic matter makes up about 4.6%. In an "extended model" which includes hot dark matter in the form of neutrinos, then if the "physical baryon density" $\grave{U}_b h^2$ is estimated at about 0.023 (this is different from the 'baryon density' $\grave{U}_b$ expressed as a fraction of the total matter/energy density, which as noted above is about 0.046), and the corresponding cold dark matter density $\grave{U}_c h^2$ is about 0.11, the corresponding neutrino density $\grave{U}_v h^2$ is estimated to be less than 0.0062. Independent lines of evidence from Type Ia supernovae and the CMB imply that the Universe today is dominated by a mysterious form of energy known as dark energy, which apparently permeates all of space.

The observations suggest 73% of the total energy density of today's Universe is in this form. When the Universe was very young, it was likely infused with dark energy, but with less space and everything closer together, gravity had the upper hand, and it was slowly braking the expansion. But eventually, after numerous billion years of expansion, the growing abundance of dark energy caused the expansion of the Universe to slowly begin to accelerate.

Dark energy in its simplest formulation takes the form of the cosmological constant term in Einstein's field equations of general relativity, but its composition and mechanism are unknown and, more generally, the details of its equation of state and relationship with the Standard Model of particle physics continue to be investigated both observationally and theoretically.

All of this cosmic evolution after the inflationary epoch can be rigorously described and modeled by the ËCDM model of cosmology, which uses the independent frameworks of quantum mechanics and Einstein's General Relativity. As noted above, there is no well-supported model describing the action prior to $10^{''15}$ seconds or so. Apparently a new unified theory of quantum gravitation is needed to break this barrier. Understanding this earliest of eras in the history of the Universe is currently one of the greatest unsolved problems in physics.

Underlying Assumptions

The Big Bang theory depends on two major assumptions: the universality of physical laws, and the cosmological principle. The cosmological principle states that on large scales the Universe is homogeneous and isotropic. These ideas were initially taken as postulates, but today there are efforts to test each of them. For example, the first assumption has been tested by observations showing that largest possible deviation of the fine structure constant over much of the age of the universe is of order 10^{-5}. Also, general relativity has passed stringent tests on the scale of the solar system and binary stars while extrapolation to cosmological scales has been validated by the empirical successes of various aspects of the Big Bang theory. If the large-scale Universe appears isotropic as viewed from Earth, the cosmological principle can be derived from the simpler Copernican principle, which states that there is no preferred (or special) observer or vantage point. To this end, the cosmological principle has been confirmed to a level of 10^{-5} via observations of the CMB. The Universe has been measured to be homogeneous on the largest scales at the 10% level.

FLRW Metric

General relativity describes spacetime by a metric, which determines the distances that separate nearby points. The points, which can be galaxies, stars, or other objects, themselves are specified using a coordinate chart or "grid" that is laid down over all spacetime. The cosmological principle implies that the metric should be homogeneous and isotropic on large scales, which uniquely singles out the Friedmann–Lemaître–Robertson–Walker metric (FLRW metric). This metric contains a scale factor, which describes how the size of the Universe changes with time. This enables a convenient choice of a coordinate system to be made, called comoving coordinates. In this coordinate system, the grid expands along with the Universe, and objects that are moving only due to the expansion of the Universe remain at fixed points on the grid. While their *coordinate* distance (comoving distance) remains constant, the *physical* distance between two such comoving points expands proportionally with the scale factor of the Universe. The Big Bang is not an explosion of matter moving outward to fill an empty universe. Instead, space itself expands with time everywhere and increases the physical distance between two comoving points. Because the FLRW metric assumes a uniform distribution of mass and energy, it applies to our Universe only on large scales—local concentrations of matter such as our galaxy are gravitationally bound and as such do not experience the large-scale expansion of space.

Horizons

An important feature of the Big Bang spacetime is the presence of horizons. Since the Universe has a finite age, and light travels at a finite speed, there may be events in the past whose light has not had time to reach us. This places a limit or a *past horizon* on the most distant objects that can be observed. Conversely, because space is expanding, and more distant objects are receding ever more quickly, light emitted by us today may never "catch up" to very distant objects. This defines a *future horizon,* which limits the events in the future that we will be able to influence. The presence of either type of horizon depends on the details of the FLRW model that describes our Universe. Our understanding of the Universe back to very early times suggests that there is a past horizon, though in practice our view is also limited by the opacity of the Universe at early times. So our view cannot extend further backward in time, though the horizon recedes in space. If the expansion of the Universe continues to accelerate, there is a future horizon as well.

Observational Evidence

The earliest and most direct kinds of observational evidence are the Hubble-type expansion seen in the redshifts of galaxies, the detailed measurements of the cosmic microwave background, the abundance of light elements, and today also the large scale distribution and apparent evolution of galaxies which are predicted to occur due to gravitational growth of structure in the standard theory. These are sometimes called *"the four pillars of the Big Bang theory".*

Hubble's Law and the Expansion of Space

Observations of distant galaxies and quasars show that these objects are redshifted—the light emitted from them has been shifted to longer wavelengths. This can be seen by taking a frequency spectrum of an object and matching the spectroscopic pattern of emission lines or absorption lines corresponding to atoms of the chemical elements interacting with the light. These redshifts are uniformly isotropic, distributed evenly among the observed objects in all directions. If the redshift is interpreted as a Doppler shift, the recessional velocity of the object can be calculated. For some galaxies, it is possible to estimate distances via the cosmic distance ladder. When the recessional velocities are plotted against these distances, a linear relationship known as Hubble's law is observed:

$$v = H_0 D,$$

where

- v is the recessional velocity of the galaxy or other distant object,
- D is the comoving distance to the object, and
- H_0 is Hubble's constant, measured to be 70.4 $70.4^{+1.3}_{-1.4}$ km/s/Mpc by the WMAP probe.

Hubble's law has two possible explanations. Either we are at the center of an explosion of galaxies—which is untenable given the Copernican Principle—or the Universe is uniformly expanding everywhere. This universal expansion was predicted from general relativity by Alexander Friedmann in 1922 and Georges Lemaître in 1927, well before Hubble made his 1929 analysis and observations, and it remains the cornerstone of the Big Bang theory as developed by Friedmann, Lemaître, Robertson and Walker.

The theory requires the relation $v = HD$ to hold at all times, where D is the comoving distance, v is the recessional velocity, and v, H, and D vary as the Universe expands (hence we write H_0 to denote the present-day Hubble "constant"). For distances much smaller than the size of the observable Universe, the Hubble redshift can be thought of as the Doppler shift corresponding to the recession velocity v. However, the redshift is not a true Doppler shift, but rather the result of the expansion of the Universe between the time the light was emitted and the time that it was detected.

That space is undergoing metric expansion is shown by direct observational evidence of the Cosmological Principle and the Copernican Principle, which together with Hubble's law have no other explanation. Astronomical redshifts are extremely isotropic and homogenous, supporting the Cosmological Principle that the Universe looks the same in all directions, along with much other evidence.

If the redshifts were the result of an explosion from a center distant from us, they would not be so similar in different directions. Measurements of the effects of the cosmic microwave background radiation on the dynamics of distant astrophysical systems in 2000 proved the Copernican Principle, that the Earth is not in a central position, on a cosmological scale. Radiation from the Big Bang was demonstrably warmer at earlier times throughout the Universe. Uniform cooling of the cosmic microwave background over billions of years is explainable only if the Universe is experiencing a metric expansion, and excludes the possibility that we are near the unique center of an explosion.

Cosmic Microwave Background Radiation

During the first few days of the Universe, the Universe was in full thermal equilibrium, with photons being continually emitted and absorbed, giving the radiation a blackbody spectrum. As the Universe expanded, it cooled to a temperature at which photons could no longer be created or destroyed. The temperature was still high enough for electrons and nuclei to remain unbound, however, and photons were constantly "reflected" from these free electrons through a process called Thomson scattering. Because of this repeated scattering, the early Universe was opaque to light.

When the temperature fell to a few thousand Kelvin, electrons and nuclei began to combine to form atoms, a process known as recombination. Since photons scatter infrequently from neutral atoms, radiation decoupled from matter when nearly all the electrons had recombined, at the *epoch of last scattering*, 379,000 years after the Big Bang. These photons make up the CMB that is observed today, and the observed pattern of fluctuations in the CMB is a direct picture of the Universe at this early epoch. The energy of photons was subsequently redshifted by the expansion of the Universe, which preserved the blackbody spectrum but caused its temperature to fall, meaning that the photons now fall into the microwave region of the electromagnetic spectrum. The radiation is thought to be observable at every point in the Universe, and comes from all directions with (almost) the same intensity.

In 1964, Arno Penzias and Robert Wilson accidentally discovered the cosmic background radiation while conducting diagnostic observations using a new microwave receiver owned by Bell Laboratories. Their discovery provided substantial confirmation of the general CMB predictions—the radiation was found to be isotropic and consistent with a blackbody spectrum of about 3 K—and it pitched the balance of opinion in favour of the Big Bang hypothesis. Penzias and Wilson were awarded a Nobel Prize for their discovery.

In 1989, NASA launched the Cosmic Background Explorer satellite (COBE), and the initial findings, released in 1990, were consistent with the Big Bang's predictions regarding the CMB. COBE found a residual temperature of 2.726 K and in 1992 detected for the first time the fluctuations (anisotropies) in the CMB, at a level of about one part in 10^5. John C. Mather and George Smoot were awarded Nobels for their leadership in this work.

During the following decade, CMB anisotropies were further investigated by a large number of ground-based and balloon experiments. In 2000–2001, several experiments, most notably BOOMERanG, found the Universe to be almost spatially flat by measuring the typical angular size (the size on the sky) of the anisotropies. In early 2003, the first results of the Wilkinson Microwave Anisotropy Probe (WMAP) were released, yielding what were at the time the most accurate values for some of the cosmological parameters. This spacecraft also disproved several specific cosmic inflation models, but the results were consistent with the inflation theory in general, it confirms too that a sea of cosmic neutrinos permeates the Universe, a clear evidence that the first stars took more than a half-billion years to create a cosmic fog.

A new space probe named Planck, with goals similar to WMAP, was launched in May 2009. It is anticipated to soon provide even more accurate measurements of the CMB anisotropies. Many other ground- and balloon-based experiments are also currently running; see Cosmic microwave background experiments. The background radiation is exceptionally smooth, which presented a problem in that conventional expansion would mean that photons coming from opposite directions in the sky were coming from regions that had never been in contact with each other. The leading explanation for this far reaching equilibrium is that the Universe had a brief period of rapid exponential expansion, called inflation. This would have the effect of driving apart regions that had been in equilibrium, so that all the observable Universe was from the same equilibrated region.

Abundance of Primordial Elements

Using the Big Bang model it is possible to calculate the concentration of helium-4, helium-3, deuterium and lithium-7 in the Universe as ratios to the amount of ordinary hydrogen, H. All the abundances depend on a single parameter, the ratio of photons to baryons, which itself can be calculated independently from the detailed structure of CMB fluctuations.

The ratios predicted (by mass, not by number) are about 0.25 for 4 He/H, about 10^{-3} for ^{2}H/H, about 10^{-4} for ^{3}He/H and about 10^{-9} for ^{7}Li/H. The measured abundances all agree at least roughly with those predicted from a single value of the baryon-to-photon ratio. The agreement is excellent for deuterium, close but formally discrepant for ^{4}He, and a factor of two off for ^{7}Li; in the latter two cases there are substantial systematic uncertainties. Nonetheless, the general consistency with abundances predicted by BBN is strong evidence for the Big Bang, as the theory is the only known explanation for the relative abundances of light elements, and it is virtually impossible to "tune" the Big Bang to produce much more or less than 20–30% helium. Indeed there is no obvious reason outside of the Big Bang that, for example, the young Universe (i.e., before star formation, as determined by studying matter supposedly free of stellar nucleosynthesis products) should have more helium than deuterium or more deuterium than ^{3}He, and in constant ratios, too.

Galactic Evolution and Distribution

Detailed observations of the morphology and distribution of galaxies and quasars provide strong evidence for the Big Bang. A combination of observations and theory suggest that the first quasars and galaxies formed about a billion years after the Big Bang, and since then larger structures have been forming, such as galaxy clusters and superclusters. Populations of stars have been aging and evolving, so that distant galaxies (which are observed as they were in the early Universe) appear very different from nearby galaxies (observed in a more recent state). Moreover, galaxies that formed relatively recently appear markedly different from galaxies formed at similar distances but shortly after the Big Bang. These observations are strong arguments against the steady-state model. Observations of star formation, galaxy and quasar distributions and larger structures agree well with Big Bang simulations of the formation of structure in the Universe and are helping to complete details of the theory.

Other Lines of Evidence

After some controversy, the age of Universe as estimated from the Hubble expansion and the CMB is now in good agreement with (i.e., slightly larger than) the ages of the oldest stars, both as measured by applying the theory of stellar evolution to globular clusters and through radiometric dating of individual Population II stars. The prediction that the CMB temperature was higher in the past has been experimentally supported by observations of temperature-sensitive emission lines in gas clouds at high redshift. This prediction also implies that the amplitude of the Sunyaev–Zel'dovich effect in clusters of galaxies does not depend directly on redshift; this seems to be roughly true, but unfortunately the amplitude does depend on cluster properties which do change substantially over cosmic time, so a precise test is impossible.

Features, Issues and Problems

While scientists now prefer the Big Bang model over other cosmological models, the scientific community was once divided between supporters of the Big Bang and those of alternative cosmological models. Throughout the historical development of the subject, problems with the Big Bang theory were posed in the context of a scientific controversy regarding which model could best describe the cosmological observations. With the overwhelming consensus in the community today supporting the Big Bang model, many of these problems are remembered as being mainly of historical interest; the solutions to them have been obtained either through modifications to the theory or as the result of better observations.

The core ideas of the Big Bang—the expansion, the early hot state, the formation of helium, the formation of galaxies—are derived from many observations that are independent from any cosmological model; these include the abundance of light elements, the cosmic microwave background, large scale structure, and the Hubble diagram for Type Ia supernovae. Precise modern models of the Big Bang appeal to various exotic physical phenomena that have not been observed in terrestrial laboratory experiments or incorporated into the Standard Model of particle physics.

Of these features, dark matter is currently the subject to the most active laboratory investigations. Remaining issues, such as the cuspy halo problem and the dwarf galaxy problem of cold dark matter, are not fatal to the dark matter explanation as solutions to such problems exist which involve only further refinements of the theory. Dark energy is also an area of intense interest for scientists, but it is not clear whether direct detection of dark energy will be possible. On the other hand, inflation and baryogenesis remain somewhat more speculative features of current Big Bang models: they explain important features of the early universe, but could be replaced by alternative ideas without affecting the rest of the theory. Discovering the correct explanations for such phenomena are some of the remaining unsolved problems in physics.

Horizon Problem

The horizon problem results from the premise that information cannot travel faster than light. In a Universe of finite age, this sets a limit—the particle horizon—on the separation of any two regions of space that are in causal contact. The observed isotropy of the CMB is problematic in this regard: if the Universe had been dominated by radiation or matter at all times up to the epoch of last scattering, the particle horizon at that time would correspond to about 2 degrees on the sky. There would then be no mechanism to cause wider regions to have the same temperature. A resolution to this apparent inconsistency is offered by inflationary theory in which a homogeneous and isotropic scalar energy field dominates the Universe at some very early period (before baryogenesis).

During inflation, the Universe undergoes exponential expansion, and the particle horizon expands much more rapidly than previously assumed, so that regions presently on opposite sides of the observable Universe are well inside each other's particle horizon. The observed isotropy of the CMB then follows from the fact that this larger region was in causal contact before the beginning of inflation.

Heisenberg's uncertainty principle predicts that during the inflationary phase there would be quantum thermal fluctuations, which would be magnified to cosmic scale. These fluctuations serve as the seeds of all current structure in the Universe. Inflation predicts that the primordial fluctuations are nearly scale invariant and Gaussian, which has been accurately confirmed by measurements of the CMB. If inflation occurred, exponential expansion would push large regions of space well beyond our observable horizon.

Flatness/Oldness Problem

The flatness problem (also known as the oldness problem) is an observational problem associated with a Friedmann–Lemaître–Robertson–Walker metric.

The Universe may have positive, negative or zero spatial curvature depending on its total energy density. Curvature is negative if its density is less than the critical density, positive if greater, and zero at the critical density, in which case space is said to be *flat*. The problem is that any small departure from the critical density grows with time, and yet the Universe today remains very close to flat.

Given that a natural timescale for departure from flatness might be the Planck time, $10^{"43}$ seconds, the fact that the Universe has reached neither a Heat Death nor a Big Crunch after billions of years requires some explanation. For instance, even at the relatively late age of a few minutes (the time of nucleosynthesis), the Universe density must have been within one part in 10^{14} of its critical value, or it would not exist as it does today.

A resolution to this problem is offered by inflationary theory. During the inflationary period, spacetime expanded to such an extent that its curvature would have been smoothed out. Thus, it is theorized that inflation drove the Universe to a very nearly spatially flat state, with almost exactly the critical density.

Magnetic Monopoles

The magnetic monopole objection was raised in the late 1970s. Grand unification theories predicted topological defects in space that would manifest as magnetic monopoles. These objects would be produced efficiently in the hot early Universe, resulting in a density much higher than is consistent with observations, given that searches have never found any monopoles. This problem is also resolved by cosmic inflation, which removes all point defects from the observable Universe in the same way that it drives the geometry to flatness. A resolution to the horizon, flatness, and magnetic monopole problems alternative to cosmic inflation is offered by the Weyl curvature hypothesis.

Baryon Asymmetry

It is not yet understood why the Universe has more matter than antimatter. It is generally assumed that when the Universe was young and very hot, it was in statistical equilibrium and contained equal numbers of baryons and antibaryons. However, observations suggest that the Universe, including its most distant parts, is made almost entirely of matter. An unknown process called "baryogenesis" created the asymmetry. For baryogenesis to occur, the Sakharov conditions must be satisfied. These require that baryon number is not conserved, that C-symmetry and CP-symmetry are violated and that the Universe depart from thermodynamic equilibrium. All these conditions occur in the Standard Model, but the effect is not strong enough to explain the present baryon asymmetry.

Globular Cluster Age

In the mid-1990s, observations of globular clusters appeared to be inconsistent with the Big Bang. Computer simulations that matched the observations of the stellar populations of globular clusters suggested that they were about 15 billion years old, which conflicted with the 13.7 billion year age of the Universe. This issue was generally resolved in the late 1990s when new computer simulations, which included the effects of mass loss due to stellar winds, indicated a much younger age for globular clusters. There still remain some questions as to how accurately the ages of the clusters are measured, but it is clear that these objects are some of the oldest in the Universe.

Dark Matter

During the 1970s and 1980s, various observations showed that there is not sufficient visible matter in the Universe to account for the apparent strength of gravitational forces within and between galaxies. This led to the idea that up to 90% of the matter in the Universe is dark matter that does not emit light or interact with normal baryonic matter.

In addition, the assumption that the Universe is mostly normal matter led to predictions that were strongly inconsistent with observations. In particular, the Universe today is far more lumpy and contains far less deuterium than can be accounted for without dark matter. While dark matter was initially controversial, it is now indicated by numerous observations: the anisotropies in the CMB, galaxy cluster velocity dispersions, large-scale structure distributions, gravitational lensing studies, and X-ray measurements of galaxy clusters. The evidence for dark matter comes from its gravitational influence on other matter, and no dark matter particles have been observed in laboratories. Many particle physics candidates for dark matter have been proposed, and several projects to detect them directly are underway.

Dark Energy

Measurements of the redshift–magnitude relation for type Ia supernovae indicate that the expansion of the Universe has been accelerating since the Universe was about half its present age. To explain this acceleration, general relativity requires that much of the energy in the Universe consists of a component with large negative pressure, dubbed "dark energy". Dark energy is indicated by several other lines of evidence.

Measurements of the cosmic microwave background indicate that the Universe is very nearly spatially flat, and therefore according to general relativity the Universe must have almost exactly the critical density of mass/energy. But the mass density of the Universe can be measured from its gravitational clustering, and is found to have only about 30% of the critical density. Since dark energy does not cluster in the usual way it is the best explanation for the "missing" energy density. Dark energy is also required by two geometrical measures of the overall curvature of the Universe, one using the frequency of gravitational lenses, and the other using the characteristic pattern of the large-scale structure as a cosmic ruler. Negative pressure is a property of vacuum energy, but the exact nature of dark energy remains one of the great mysteries of the Big Bang. Possible candidates include a cosmological constant and quintessence. Results from the WMAP team in 2008, which combined data from the CMB and other sources, indicate that the Universe today is 73% dark energy, 23% dark matter, 4.6% regular matter and less than 1% neutrinos.

The energy density in matter decreases with the expansion of the Universe, but the dark energy density remains constant (or nearly so) as the Universe expands. Therefore matter made up a larger fraction of the total energy of the Universe in the past than it does today, but its fractional contribution will fall in the far future as dark energy becomes even more dominant.

In the ËCDM, the best current model of the Big Bang, dark energy is explained by the presence of a cosmological constant in the general theory of relativity. However, the size of the constant that properly explains dark energy is surprisingly small relative to naive estimates based on ideas about quantum gravity. Distinguishing between the cosmological constant and other explanations of dark energy is an active area of current research.

The Future According to the Big Bang Theory

Before observations of dark energy, cosmologists considered two scenarios for the future of the Universe. If the mass density of the Universe were greater than the critical density, then the Universe would reach a maximum size and then begin to collapse. It would become denser and hotter again, ending with a state that was similar to that in which it started—a Big Crunch.

Alternatively, if the density in the Universe were equal to or below the critical density, the expansion would slow down, but never stop. Star formation would cease as all the interstellar gas in each galaxy is consumed; stars would burn out leaving white dwarfs, neutron stars, and black holes. Very gradually, collisions between these would result in mass accumulating into larger and larger black holes. The average temperature of the Universe would asymptotically approach absolute zero—a Big Freeze. Moreover, if the proton were unstable, then baryonic matter would disappear, leaving only radiation and black holes. Eventually, black holes would evaporate by emitting Hawking radiation. The entropy of the Universe would increase to the point where no organized form of energy could be extracted from it, a scenario known as heat death. Modern observations of accelerated expansion imply that more and more of the currently visible Universe will pass beyond our event horizon and out of contact with us. The eventual result is not known.

The ËCDM model of the Universe contains dark energy in the form of a cosmological constant. This theory suggests that only gravitationally bound systems, such as galaxies, would remain together, and they too would be subject to heat death, as the Universe expands and cools. Other explanations of dark energy—so-called phantom energy theories—suggest that ultimately galaxy clusters, stars, planets, atoms, nuclei and matter itself will be torn apart by the ever-increasing expansion in a so-called Big Rip.

Speculative Physics Beyond Big Bang Theory

While the Big Bang model is well established in cosmology, it is likely to be refined in the future. Little is known about the earliest moments of the Universe's history. The Penrose–Hawking singularity theorems require the existence of a singularity at the beginning of cosmic time. However, these theorems assume that general relativity is correct, but general relativity must break down before the Universe reaches the Planck temperature, and a correct treatment of quantum gravity may avoid the singularity. Some proposals, each of which entails untested hypotheses, are:

- models including the Hartle–Hawking no-boundary condition in which the whole of space-time is finite; the Big Bang does represent the limit of time, but without the need for a singularity.
- brane cosmology models in which inflation is due to the movement of branes in string theory; the pre-big bang model; the ekpyrotic model, in which the Big Bang is the result of a collision between branes; and the cyclic model, a variant of the ekpyrotic model in which collisions occur periodically. In the latter model, the Big Bang was preceded by a Big Crunch and the Universe endlessly cycles from one process to the other.
- chaotic inflation, in which universal inflation ends locally here and there in a random fashion, each end-point leading to a *bubble universe* expanding from its own big bang.

Proposals in the last two categories see the Big Bang as an event in a much larger and older Universe, or multiverse, and not the literal beginning.

Religious Interpretations

The Big Bang is a scientific theory, and as such is dependent on its agreement with observations. But as a theory which addresses the origins of reality, it has always carried theological and philosophical implications, most notably, the concept of creation *ex nihilo*. In the 1920s and 1930s almost every major cosmologist preferred an eternal steady state Universe, and several complained that the beginning of time implied by the Big Bang imported religious concepts into physics;

this objection was later repeated by supporters of the steady state theory. This perception was enhanced by the fact that the originator of the Big Bang theory, Monsignor Georges Lemaître, was a Roman Catholic Christian priest.

Pope Pius XII, declared at the November 22, 1951 opening meeting of the Pontifical Academy of Sciences that the Big Bang theory accorded with the Catholic concept of creation. Conservative Protestant Christian denominations have also welcomed the Big Bang theory as supporting a historical interpretation of the doctrine of creation. Since the acceptance of the Big Bang as the dominant physical cosmological paradigm, there have been a variety of reactions by religious groups as to its implications for their respective religious cosmologies. Some accept the scientific evidence at face value, while others seek to reconcile the Big Bang with their religious tenets, and others completely reject or ignore the evidence for the Big Bang theory.

Nuclear Reactor Physics

Most nuclear reactors use a chain reaction to induce a controlled rate of nuclear fission in fissile material, releasing both energy and free neutrons. A reactor consists of an assembly of nuclear fuel (a reactor core), usually surrounded by a neutron moderator such as regular water, heavy water, graphite, or zirconium hydride, and fitted with mechanisms such as control rods that control the rate of the reaction. Nuclear reactor physics is the branch of science that deals with the study and application of chain reaction to induce controlled rate of fission for energy in reactors. The physics of nuclear fission has several quirks that affect the design and behaviour of nuclear reactors.

Starter Sources

The mere fact that an assembly is supercritical does not guarantee that it contains any free neutrons at all. At least one neutron is required to "strike" a chain reaction, and if the spontaneous fission rate is sufficiently low it may take a long time (in ^{235}U reactors, as long as many minutes) before a chance neutron encounter starts a chain reaction even if the reactor is supercritical. Most nuclear reactors include a "starter" neutron source that ensures there are always a few free neutrons in the reactor core, so that a chain reaction will begin immediately when the core is made critical.

A common type of startup neutron source is a mixture of an alpha particle emitter such as ^{241}Am (americium-241) with a lightweight isotope such as ^{9}Be (beryllium-9). The primary sources described above have to be used with fresh reactor cores. For operational reactors, secondary sources are used; most often a combination of antimony with beryllium. Antimony becomes activated in the reactor and produces high-energy gamma photons, which produce photoneutrons from beryllium.

Uranium-235 undergoes a small rate of natural spontaneous fission, so there are always some neutrons being produced even in a fully shutdown reactor. When the control rods are withdrawn and criticality is approached the number increases because the absorption of neutrons is being progressively reduced, until at criticality the chain reaction becomes self-sustaining.

Note that while a neutron source is provided in the reactor, this is not essential to start the chain reaction, its main purpose is to give a shutdown neutron population which is detectable by instruments and so make the approach to critical more observable. The reactor will go critical at the same control rod position whether a source is loaded or not. Once the chain reaction is begun, the primary starter source may be removed from the core to prevent damage from the high neutron flux in the operating reactor core; the secondary sources usually remains in situ to provide a background reference level for control of criticality.

Subcritical Multiplication

Even in a subcritical assembly such as a shut-down reactor core, any stray neutron that happens to be present in the core (for example from spontaneous fission of the fuel, from radioactive decay of fission products, or from a neutron source) will trigger an exponentially decaying chain reaction. Although the chain reaction is not self-sustaining, it acts as a multiplier that increases the equilibrium number of neutrons in the core. This *subcritical multiplication* effect can be used in two ways: as a probe of how close a core is to criticality, and as a way to generate fission power without the risks associated with a critical mass. As a measurement technique, subcritical multiplication was used during the Manhattan Project in early experiments to determine the minimum critical masses of ^{235}U and of ^{239}Pu. It is still used today to calibrate the controls for nuclear reactors during startup, as many effects (discussed in the following sections) can change the required control settings to achieve criticality in a reactor.

As a power-generating technique, subcritical multiplication allows generation of nuclear power for fission where a critical assembly is undesirable for safety or other reasons. A subcritical assembly together with a neutron source can serve as a steady source of heat to generate power from fission. Including the effect of an external neutron source ("external" to the fission process, not physically external to the core), one can write a modified evolution equation:

$$dN / dt = \alpha N / \tau + R_{ext}$$

where R_{ext} is the rate at which the external source injects neutrons into the core. In equilibrium, the core is not changing and dN/dt is zero, so the equilibrium number of neutrons is given by:

$$N = \tau R_{ext} / (-\alpha)$$

If the core is subcritical, then α is negative so there is an equilibrium with a positive number of neutrons. If the core is close to criticality, then α is very small and thus the final number of neutrons can be made arbitrarily large.

Neutron Moderators

To improve $P_{fission}$ and enable a chain reaction, uranium-fueled reactors must include a neutron moderator that interacts with newly produced fast neutrons from fission events to reduce their kinetic energy from several MeV to several eV, making them more likely to induce fission. This is because ^{235}U is much more likely to undergo fission when struck by one of these thermal neutrons than by a freshly-produced neutron from fission. Neutron moderators are materials that interact weakly with the neutrons but absorb kinetic energy from them. Most moderators rely on either weakly bound hydrogen or a loose crystal structure of another light element such as carbon to transfer kinetic energy from the fast-moving neutrons. Hydrogen moderators include water (H_2O), heavy water(D_2O), and zirconium hydride (ZrH_2), all of which work because a hydrogen nucleus has nearly the same mass as a free neutron: neutron-H_2O or neutron-ZrH_2 impacts excite rotational modes of the molecules (spinning them around).

Deuterium nuclei (in heavy water) absorb kinetic energy less well than do light hydrogen nuclei, but they are much less likely to absorb the impacting neutron. Water or heavy water have the advantage of being transparent liquids, so that, in addition to shielding and moderating a reactor core, they permit direct viewing of the core in operation and can also serve as a working fluid for heat transfer. Crystal structure moderators rely on a floppy crystal matrix to absorb phonons from neutron-crystal impacts. Graphite is the most common example of such a moderator. It was used in Chicago Pile-1, the world's first man-made critical assembly, and was commonplace in early reactor designs including the Soviet RBMK nuclear power plants, of which the Chernobyl plant was one.

Moderators and Reactor Design

The amount and nature of neutron moderation affects reactor controllability and hence safety. Because moderators both slow and absorb neutrons, there is an optimum amount of moderator to include in a given geometry of reactor core. Less moderation reduces the effectiveness by reducing the $P_{fission}$ term in the evolution equation, and more moderation reduces the effectiveness by increasing the P_{escape} term. Most moderators become less effective with increasing temperature, so *under-moderated* reactors are stable against changes in temperature in the reactor core: if the core overheats, then the quality of the moderator is reduced and the reaction tends to slow down (there is a "negative temperature coefficient" in the reactivity of the core). Water is an extreme case: in extreme heat, it can boil, producing effective voids in the reactor core without destroying the physical structure of the core; this tends to shut down the reaction and reduce the possibility of a fuel meltdown. *Over-moderated* reactors are unstable against changes in temperature (there is a "positive temperature coefficient" in the reactivity of the core), and so are less inherently safe than under-moderated cores. Some reactors use a combination of moderator materials. For example, TRIGA type research reactors use ZrH_2 moderator mixed with the ^{235}U fuel, an H_2O-filled core, and C (graphite) moderator and reflector blocks around the periphery of the core.

Delayed Neutrons and Controllability

Fission reactions and subsequent neutron escape happen very quickly; this is important for nuclear weapons, where the object is to make a nuclear core release as much energy as possible before it physically explodes. Most neutrons emitted by fission events are prompt: they are emitted essentially instantaneously. Once emitted, the average neutron lifetime (τ) in a typical core is on the order of a millisecond, so if the exponential factor α is as small as 0.01, then in one second the reactor power will vary by a factor of $(1+0.01)^{1000}$, or more than ten thousand.

Nuclear weapons are engineered to maximize the power growth rate, with lifetimes well under a millisecond and exponential factors close to 2; but such rapid variation would render it practically impossible to control the reaction rates in a nuclear reactor. Fortunately, the *effective* neutron lifetime is much longer than the average lifetime of a single neutron in the core. About 0.65% of the neutrons produced by ^{235}U fission, and about 0.75% of the neutrons produced by ^{239}Pu fission, are not produced immediately, but rather are emitted by radioactive decay of fission products, with an average lifetime of about 15 seconds.

These delayed neutrons increase the effective average lifetime of neutrons in the core, to nearly 0.1 seconds, so that a core with α of 0.01 would increase in one second by only a factor of $(1+0.01)^{10}$, or about 1.1 — a 10% increase.

This is a controllable rate of change. Most nuclear reactors are hence operated in a *prompt subcritical, delayed critical* condition: the prompt neutrons alone are not sufficient to sustain a chain reaction, but the delayed neutrons make up the small difference required to keep the reaction going.

This has effects on how reactors are controlled: when a small amount of control rod is slid into or out of the reactor core, the power level changes at first very rapidly due to *prompt subcritical multiplication* and then more gradually, following the exponential growth or decay curve of the delayed critical reaction. Further, *increases* in reactor power can be performed at any desired rate simply by pulling out a sufficient length of control rod — but *decreases* are limited in speed, because even if the reactor is taken deeply subcritical, the delayed neutrons are produced by ordinary radioactive decay of fission products and that decay cannot be hastened.

Reactor Poisons

Any element that strongly absorbs neutrons is called a reactor poison, because it tends to shut down (poison) an ongoing fission chain reaction. Some reactor poisons are deliberately inserted into fission reactor cores to control the reaction; boron or cadmium control rods are the best example. Many reactor poisons are produced by the fission process itself, and buildup of neutron-absorbing fission products affects both the fuel economics and the controllability of nuclear reactors.

Long-lived Poisons and Fuel Reprocessing

In practice, buildup of reactor poisons in nuclear fuel is what determines the lifetime of nuclear fuel in a reactor: long before all possible fissions have taken place, buildup of long-lived neutron absorbing fission products damps out the chain reaction. This is the reason that nuclear reprocessing is a useful activity: spent nuclear fuel contains about 99% of the original fissionable material present in newly manufactured nuclear fuel.

Chemical separation of the fission products restores the nuclear fuel so that it can be used again. Nuclear reprocessing is useful economically because chemical separation is much simpler to accomplish than the difficult isotope separation required to prepare nuclear fuel from natural uranium ore, so that in principle chemical separation yields more generated energy for less effort than mining, purifying, and isotopically separating new uranium ore. In practice, both the difficulty of handling the highly radioactive fission products and other political concerns make fuel reprocessing a contentious subject. One such concern is the fact that spent uranium nuclear fuel contains significant quantities of ^{239}Pu, a prime ingredient in nuclear weapons.

Short-lived Poisons and Controllability

Short-lived reactor poisons in fission products strongly affect how nuclear reactors can operate. Unstable fission product nuclei transmute into many different elements (*secondary fission products*) as they undergo a decay chain to a stable isotope. The most important such element is xenon, because the isotope ^{135}Xe, a secondary fission product with a half-life of about 9 hours, is an extremely strong neutron absorber. In an operating reactor, each nucleus of ^{135}Xe is destroyed by neutron capture almost as soon as it is created, so that there is no buildup in the core.

However, when a reactor shuts down, the level of ^{135}Xe builds up in the core for about 9 hours before beginning to decay. The result is that, about 6-8 hours after a reactor is shut down, it can become physically impossible to restart the chain reaction until the ^{135}Xe has had a chance to decay over the next several hours. This temporary state, which may last several days and prevent restart, is called the iodine pit or xenon-poisoning.

It is one reason why nuclear power reactors are usually operated at an even power level around the clock. ^{135}Xe buildup in a reactor core makes it extremely dangerous to operate the reactor a few hours after it has been shut down. Because the ^{135}Xe absorbs neutrons strongly, starting a reactor in a high-Xe condition requires pulling the control rods out of the core much farther than normal.

However, if the reactor does achieve criticality, then the neutron flux in the core becomes high and ^{135}Xe is destroyed rapidly — this has the same effect as very rapidly removing a great length of control rod from the core, and can cause the reaction to grow too rapidly or even become prompt critical. ^{135}Xe played a large part in the Chernobyl accident: about eight hours after a scheduled maintenance shutdown, workers tried to bring the reactor to a zero power critical condition to test a control circuit. Since the core was loaded with ^{135}Xe from the previous day's power generation, it was necessary to withdraw more control rods to achieve this.

As a result, the overdriven reaction grew rapidly and uncontrollably, leading to steam explosion in the core, fire in the graphite moderator, and violent destruction of the facility.

Uranium Enrichment

While many fissionable isotopes exist in nature, the only usefully fissile isotope found in any quantity is ^{235}U. About 0.7% of the uranium in most ores is the 235 isotope, and about 99.3% is the inert 238 isotope. For most uses as a nuclear fuel, uranium must be *enriched* - purified so that it contains a higher percentage of ^{235}U. Because ^{238}U absorbs fast neutrons, the critical mass needed to sustain a chain reaction increases as the ^{238}U content increases, reaching infinity at 94% ^{238}U (6% ^{235}U). Concentrations lower than 6% ^{235}U cannot go fast critical, though they are usable in a nuclear reactor with a neutron moderator. A nuclear weapon primary stage using uranium uses HEU enriched to ~90% ^{235}U, though the secondary stage often uses lower enrichments. Nuclear reactors with water moderator can operate with only moderate enrichment of ~5% ^{235}U. Nuclear reactors with heavy water moderation can operate with natural uranium, eliminating altogether the need for enrichment and preventing the fuel from being useful for nuclear weapons; the CANDU power reactors used in Canadian power plants are an example of this type. Uranium enrichment is difficult because the chemical properties of ^{235}U and ^{238}U are identical, so physical processes such as gaseous diffusion, gas centrifuge or mass spectrometry must be used for isotopic separation based on small differences in mass. Because enrichment is the main technical hurdle to production of nuclear fuel and simple nuclear weapons, enrichment technology is politically sensitive.

Oklo: A Natural Nuclear Reactor

Modern deposits of uranium contain only up to ~0.7% ^{235}U (and ~99.3% ^{238}U), which is not enough to sustain a chain reaction moderated by ordinary water. But ^{235}U has a much shorter half-life (700 million years) than ^{238}U (4.5 billion years), so in the distant past the percentage of ^{235}U was much higher. About two billion years ago, a water-saturated uranium deposit (in what is now the Oklo mine in Gabon, West Africa) underwent a naturally occurring chain reaction that was moderated by groundwater and, presumably, controlled by the negative void coefficient as the water boiled from the heat of the reaction. Uranium from the Oklo mine is about 50% depleted compared to other locations: it is only about 0.3% to 0.7% ^{235}U; and the ore contains traces of stable daughters of long-decayed fission products.

Nuclear Fission

In nuclear physics and nuclear chemistry, nuclear fission is a nuclear reaction in which the nucleus of an atom splits into smaller parts (lighter nuclei), often producing free neutrons and photons (in the form of gamma rays).

The two nuclei produced are most often of comparable size, typically with a mass ratio around 60:40. Most fissions are binary fissions, but occasionally (2 to 4 times per 1000 events), three positively-charged fragments are produced in a ternary fission. The smallest of these range in size from protons to argon nuclei. Fission is usually the product of an energetic nuclear reaction induced by a neutron, although it is occasionally seen as a form of spontaneous radioactive decay. The lack of any defined composition in the products (which are produced in a broad probabilistic manner) distinguishes fission from quantum tunnelling processes like proton emission, alpha decay and cluster decay. Fission of heavy elements is an exothermic reaction which can release large amounts of energy both as electromagnetic radiation and as kinetic energy of the fragments (heating the bulk material where fission takes place).

In order for fission to produce energy, the total binding energy of the resulting elements must be less than that of the starting element. Fission is a form of nuclear transmutation because the resulting fragments are not the same element as the original atom. Nuclear fission produces energy for nuclear power and to drive the explosion of nuclear weapons.

Both uses are possible because certain substances called nuclear fuels undergo fission when struck by fission neutrons, and in turn emit neutrons when they break apart. This makes possible a self-sustaining chain reaction that releases energy at a controlled rate in a nuclear reactor or at a very rapid uncontrolled rate in a nuclear weapon. The amount of free energy contained in nuclear fuel is millions of times the amount of free energy contained in a similar mass of chemical fuel such as gasoline, making nuclear fission a very tempting source of energy. The products of nuclear fission, however, are on average far more radioactive than the heavy elements which are normally fissioned as fuel, and remain so for significant amounts of time, giving rise to a nuclear waste problem. Concerns over nuclear waste accumulation and over the destructive potential of nuclear weapons may counterbalance the desirable qualities of fission as an energy source, and give rise to ongoing political debate over nuclear power.

Mechanics

Nuclear fission can occur without neutron bombardment, as a type of radioactive decay. This type of fission (called spontaneous fission) is rare except in a few heavy isotopes. In engineered nuclear devices, essentially all nuclear fission occurs as a "nuclear reaction" — a bombardment-driven process that results from the collision of two subatomic particles.

In nuclear reactions, a subatomic particle collides with an atomic nucleus and causes changes to it. Nuclear reactions are thus driven by the mechanics of bombardment, not by the relatively constant exponential decay and half-life characteristic of spontaneous radioactive processes. Many types of nuclear reactions are currently known. Nuclear fission differs importantly from other types of nuclear reactions, in that it can be amplified and sometimes controlled via a nuclear chain reaction. In such a reaction, free neutrons released by each fission event can trigger yet more events, which in turn release more neutrons and cause more fissions. The chemical element isotopes that can sustain a fission chain reaction are called nuclear fuels, and are said to be *fissile*.

The most common nuclear fuels are ^{235}U (the isotope of uranium with an atomic mass of 235 and of use in nuclear reactors) and ^{239}Pu (the isotope of plutonium with an atomic mass of 239). These fuels break apart into a bimodal range of chemical elements with atomic masses centering near 95 and 135 u (fission products). Most nuclear fuels undergo spontaneous fission only very slowly, decaying instead mainly via an alpha/beta decay chain over periods of millennia to eons. In a nuclear reactor or nuclear weapon, the overwhelming majority of fission events are induced by bombardment with another particle, a neutron, which is itself produced by prior fission events.

Nuclear fission is normally the result of the nuclear excitation energy produced when a fissionable nucleus captures a neutron. This energy, resulting from the neutron capture, is a result of the attractive nuclear force acting between the neutron and nucleus. It is enough to deform the nucleus into a double-lobed "drop," to the point that nuclear fragments exceed the distances at which the nuclear force can hold two groups of charged nucleons together, and when this happens, the two fragments complete their separation and then are driven further appart by their mutually repulsive charges, in a process which becomes irreversible with greater and greater distance.

The liquid drop model of the atomic nucleus predicts equal-sized fission products as a mechanical outcome of nuclear deformation. The more sophisticated nuclear shell model is needed to mechanistically explain the route to the more energetically-favourable outcome, in which one fission product is slightly smaller than the other. The most common fission process is binary fission, and it produces the fission products noted above, at 95±15 and 135±15 u.

However, the binary process happens merely because it is the most probable. In anywhere from 2 to 4 fissions per 1000 in a nuclear reactor, a process called ternary fission produces three positively charged fragments (plus neutrons) and the smallest of these may range from so small a charge and mass as a proton (Z=1), to as large a fragment as argon (Z=18).

The most common small fragments, however, are composed of 90% helium-4 nuclei with more energy than alpha particles from alpha decay (so-called "long range alphas" at ~ 16 MeV), plus helium-6 nuclei, and tritons (the nuclei of tritium). The ternary process is less common, but still ends up producing significant helium-4 and tritium gas buildup in the fuel rods of modern nucler reactors.

Energetics

Input: The fission of a heavy nucleus requires a total input energy of about 7 to 8 MeV to initially overcome the strong force which holds the nucleus into a spherical or nearly spherical shape, and from there, deform it into a two-lobed ("peanut") shape in which the lobes are able to continue to separate from each other, pushed by their mutual positive charge, in the most common process of binary fission (two positively-charged fission products + neutrons).

Once the nuclear lobes have been pushed to a critical distance, beyond which the short range strong force can no longer hold them together, the process of their separation proceeds from the energy of the (longer range) electromagnetic repulsion between the fragments.

The result is two fission fragments moving away from each other, at high energy. About 6 MeV of the fission-input energy is supplied by the simple binding of the neutron to the nucleus via the strong force—however in many fissionable isotopes, this amount of energy is not enough for fission. If no additional energy is supplied by any other mechanism, the nucleus will not fission, but will merely absorb the neutron, as happens when U-238 absorbs slow neutrons to become U-239. The remaining energy to initiate fission can be supplied by two other mechanisms: one of these is the kinetic energy of the incoming neutron, which is increasingly able to fission a fissionable heavy nucleus as it exceeds a kinetic energy of one MeV or more (so-called fast neutrons).

Such high energy neutrons are able to fission U-238 directly. However, this process cannot happen to a great extent in a nuclear reactor, as too small a fraction of the fission neutrons produced by any type of fission have enough energy to directly fission U-238. Among the heavy actinide elements, however, those isotopes that have an odd number of neutrons bind neutrons with an additional 1 to 2 MeV of energy, which is made available as a result of the mechanism of neutron pairing effects. This extra energy results from the Pauli exclusion principle allowing an extra neutron to occupy the same nuclear orbital as the last neutron in the nucleus, so that the two form a pair. In such isotopes, therefore, no neutron kinetic energy is needed, for all the necessary energy is supplied by absorption of any neutron, either of the slow or fast variety (the former are used in nuclear reactors, and the latter are used in weapons). As noted above, the subgroup of fissionable elements that may be fissioned with their own fission neutrons, are termed fissile. Examples of fissile isotopes are U-235 and plutonium-239.

Output: Typical fission events release about two hundred million eV (200 MeV) of energy for each fission event. By contrast, most chemical oxidation reactions (such as burning coal or TNT) release at most a few eV per event, so nuclear fuel contains at least ten million times more usable energy per unit mass than does chemical fuel. The energy of nuclear fission is released as kinetic energy of the fission products and fragments, and as electromagnetic radiation in the form of gamma rays; in a nuclear reactor, the energy is converted to heat as the particles and gamma rays collide with the atoms that make up the reactor and its working fluid, usually water or occasionally heavy water.

When a uranium nucleus fissions into two daughter nuclei fragments, an energy of ~200 MeV is released. For uranium-235 (total mean fission energy 202.5 MeV), typically ~169 MeV appears as the kinetic energy of the daughter nuclei, which fly apart at about 3% of the speed of light, due to Coulomb repulsion. Also, an average of 2.5 neutrons are emitted with a kinetic energy of ~2 MeV each (total of 4.8 MeV). The fission reaction also releases ~7 MeV in prompt gamma ray photons.

The latter figure means that a nuclear fission explosion or criticality accident emits about 3.5% of its energy as gamma rays, less than 2.5% of its energy as fast neutrons (total ~ 6%), and the rest as kinetic energy of fission fragments ("heat"). In an atomic bomb, this heat may serve to raise the temperature of the bomb core to 100 million kelvin and cause secondary emission of soft X-rays, which convert some of this energy to ionizing radiation.

However, in nuclear generators, the fission fragment kinetic energy remains as low-temperature heat which causes little or no ionization. So-called neutron bombs (enhanced radiation weapons) have been constructed which release a larger fraction of their energy as ionizing radiation (specifically, neutrons), but these are all thermonuclear devices which rely on the nuclear fusion stage to produce the extra radiation. The energy dynamics of pure fission bombs always remain at about 6% yield of the total in radiation, as a prompt result of fission. The total *prompt fission* energy amounts to about 181 MeV, or ~ 89% of the total energy which is eventually released by fission over time.

The remaining ~ 11% is released in beta decays which have various half-lives, but begin as a process in the fission products immediately; and in delayed gamma emissions associated with these beta decays. For example, in uranium-235 this delayed energy is divided into about 6.5 MeV in betas, 8.8 MeV in antineutrinos (released at the same time as the betas), and finally, an additional 6.3 MeV in delayed gamma emission from the excited beta-decay products (for a mean total of ~10 gamma ray emissions per fission, in all). Thus, an additional 6% of the total energy of fission is also released eventually as non-prompt ionizing radiation, and this is about evenly divided between gamma and beta ray energy.

The remainder is antineutrinos. The 8.8 MeV/202.5 MeV = 4.3% of the energy which is released as antineutrinos is not captured by the reactor material as heat, and escapes directly through all materials (including the Earth) at nearly the speed of light, and into interplanetary space (the amount absorbed is miniscule). Neutrino radiation is ordinarily not classed as ionizing radiation, because it is not absorbed and therefore does not produce effects. Almost all of the rest of the radiation (beta and gamma radiation) is eventually converted to heat in a reactor core or its shielding.

Some processes involving neutrons are notable for absorbing or finally yielding energy — for example neutron kinetic energy does not yield heat immediately if the neutron is captured by a uranium-238 atom to breed plutonium-239, but this energy is emitted if the plutonium-239 is later fissioned. On the other hand, so-called delayed neutrons emitted as radioactive decay products with half-lives up to several minutes, from fission-daughters, are very important to reactor control, because they give a characteristic "reaction" time for the total nuclear reaction to double in size, if the reaction is run in a "delayed-critical" zone which deliberately relies on these neutrons for a supercritical chain-reaction (one in which each fission cycle yields more neutrons than it absorbs).

Without their existence, the nuclear chain-reaction would be prompt critical and increase in size faster than it could be controlled by human intervention. In this case, the first experimental atomic reactors would have run away to a dangerous and messy "prompt critical reaction" before their operators could have manually shut them down (for this reason, designer Enrico Fermi included radiation-counter-triggered control rods, suspended by electromagnets, which could automatically drop into the center of Chicago Pile-1). If these delayed neutrons are captured without producing fissions, they produce heat as well.

Product Nuclei and Binding Energy

In fission there is a preference to yield fragments with even proton numbers, which is called the odd-even effect on the fragments charge distribution. However, no odd-even effect is observed on fragment mass number distribution. This result is attributed to nucleon pair breaking. In nuclear fission events the nuclei may break into any combination of lighter nuclei, but the most common event is not fission to equal mass nuclei of about mass 120; the most common event (depending on isotope and process) is a slightly unequal fission in which one daughter nucleus has a mass of about 90 to 100u and the other the remaining 130 to 140 u. Unequal fissions are energetically more favourable because this allows one product to be closer to the energetic minimum near mass 60 u (only a quarter of the average fissionable mass), while the other nucleus with mass 135 u is still not far out of the range of the most tightly bound nuclei (another statement of this, is that the atomic binding energy curve is slightly steeper to the left of mass 120 u than to the right of it).

Origin of the Active Energy and the Curve of Binding Energy

Nuclear fission of heavy elements produces energy because the specific binding energy (binding energy per mass) of intermediate-mass nuclei with atomic numbers and atomic masses close to ^{62}Ni and ^{56}Fe is greater than the nucleon-specific binding energy of very heavy nuclei, so that energy is released when heavy nuclei are broken apart. The total rest masses of the fission products (Mp) from a single reaction is less than the mass of the original fuel nucleus (M). The excess mass Äm = M – Mp is the invariant mass of the energy that is released as photons (gamma rays) and kinetic energy of the fission fragments, according to the mass-energy equivalence formula $E = mc^2$.

The variation in specific binding energy with atomic number is due to the interplay of the two fundamental forces acting on the component nucleons (protons and neutrons) that make up the nucleus. Nuclei are bound by an attractive nuclear force between nucleons, which overcomes the electrostatic repulsion between protons. However, the nuclear force acts only over relatively short ranges (a few nucleon diameters), since it follows an exponentially decaying Yukawa potential which makes it insignificant at longer distances. The electrostatic repulsion is of longer range, since it decays by an inverse-square rule, so that nuclei larger than about 12 nucleons in diameter reach a point that the total electrostatic repulsion overcomes the nuclear force and causes them to be spontaneously unstable.

For the same reason, larger nuclei (more than about eight nucleons in diameter) are less tightly bound per unit mass than are smaller nuclei; breaking a large nucleus into two or more intermediate-sized nuclei, releases energy. The origin of this energy is the nuclear force, which intermediate-sized nuclei allows to act more efficiently, because each nucleon has more neighbors which are within the short range attraction of this force. Thus less energy is needed in the smaller nuclei and the difference to the state before is set free.

Also because of the short range of the strong binding force, large stable nuclei must contain proportionally more neutrons than do the lightest elements, which are most stable with a 1 to 1 ratio of protons and neutrons. Nuclei which have more than 20 protons cannot be stable unless they have more than an equal number of neutrons. Extra neutrons stabilize heavy elements because they add to strong-force binding (which acts between all nucleons), without adding to proton–proton repulsion.

Fission products have, on average, about the same ratio of neutrons and protons as their parent nucleus, and are therefore usually unstable to beta decay (which changes neutrons to protons) because they have proportionally too many neutrons compared to stable isotopes of similar mass. This tendency for fission product nuclei to beta-decay is the fundamental cause of the problem of radioactive high level waste from nuclear reactors. Fission products tend to be beta emitters, emitting fast-moving electrons to conserve electric charge, as excess neutrons convert to protons in the fission-product atoms. See Fission products (by element) for a description of fission products sorted by element.

Chain Reactions

Several heavy elements, such as uranium, thorium, and plutonium, undergo both spontaneous fission, a form of radioactive decay and *induced fission*, a form of nuclear reaction. Elemental isotopes that undergo induced fission when struck by a free neutron are called fissionable; isotopes that undergo fission when struck by a thermal, slow moving neutron are also called fissile. A few particularly fissile and readily obtainable isotopes (notably ^{235}U and ^{239}Pu) are called nuclear fuels because they can sustain a chain reaction and can be obtained in large enough quantities to be useful. All fissionable and fissile isotopes undergo a small amount of spontaneous fission which releases a few free neutrons into any sample of nuclear fuel. Such neutrons would escape rapidly from the fuel and become a free neutron, with a mean lifetime of about 15 minutes before decaying to protons and beta particles.

However, neutrons almost invariably impact and are absorbed by other nuclei in the vicinity long before this happens (newly-created fission neutrons move at about 7% of the speed of light, and even moderated neutrons move at about 8 times the speed of sound). Some neutrons will impact fuel nuclei and induce further fissions, releasing yet more neutrons. If enough nuclear fuel is assembled in one place, or if the escaping neutrons are sufficiently contained, then these freshly emitted neutrons outnumber the neutrons that escape from the assembly, and a *sustained nuclear chain reaction* will take place.

An assembly that supports a sustained nuclear chain reaction is called a critical assembly or, if the assembly is almost entirely made of a nuclear fuel, a critical mass. The word "critical" refers to a cusp in the behaviour of the differential equation that governs the number of free neutrons present in the fuel: if less than a critical mass is present, then the amount of neutrons is determined by radioactive decay, but if a critical mass or more is present, then the amount of neutrons is controlled instead by the physics of the chain reaction. The actual mass of a *critical mass* of nuclear fuel depends strongly on the geometry and surrounding materials.

Not all fissionable isotopes can sustain a chain reaction. For example, ^{238}U, the most abundant form of uranium, is fissionable but not fissile: it undergoes induced fission when impacted by an energetic neutron with over 1 MeV of kinetic energy. But too few of the neutrons produced by ^{238}U fission are energetic enough to induce further fissions in ^{238}U, so no chain reaction is possible with this isotope. Instead, bombarding ^{238}U with slow neutrons causes it to absorb them (becoming ^{239}U) and decay by beta emission to ^{239}Np which then decays again by the same process to ^{239}Pu; that process is used to manufacture ^{239}Pu in breeder reactors.

In-situ plutonium production also contributes to the neutron chain reaction in other types of reactors after sufficient plutonium-239 has been produced, since plutonium-239 is also a fissile element which serves as fuel. It is estimated that up to half of the power produced by a standard "non-breeder" reactor is produced by the fission of plutonium-239 produced in place, over the total life-cycle of a fuel load. Fissionable, non-fissile isotopes can be used as fission energy source even without a chain reaction.

Bombarding ^{238}U with fast neutrons induces fissions, releasing energy as long as the external neutron source is present. This is an important effect in all reactors where fast neutrons from the fissile isotope can cause the fission of nearby ^{238}U nuclei, which means that some small part of the ^{238}U is "burned-up" in all nuclear fuels, especially in fast breeder reactors that operate with higher-energy neutrons. That same fast-fission effect is used to augment the energy released by modern thermonuclear weapons, by jacketing the weapon with ^{238}U to react with neutrons released by nuclear fusion at the center of the device.

Fission Reactors

Critical fission reactors are the most common type of nuclear reactor. In a critical fission reactor, neutrons produced by fission of fuel atoms are used to induce yet more fissions, to sustain a controllable amount of energy release. Devices that produce engineered but non-self-sustaining fission reactions are subcritical fission reactors.

Such devices use radioactive decay or particle accelerators to trigger fissions. Critical fission reactors are built for three primary purposes, which typically involve different engineering trade-offs to take advantage of either the heat or the neutrons produced by the fission chain reaction:

- *power reactors* are intended to produce heat for nulear power, either as part of a generating station or a local power system such as a nuclear submarine.

- *research reactors* are intended to produce neutrons and/or activate radioactive sources for scientific, medical, engineering, or other research purposes.
- *breeder reactors* are intended to produce nuclear fuels in bulk from more abundant isotopes. The better known fast breeder reactor makes ^{239}Pu (a nuclear fuel) from the naturally very abundant ^{238}U (not a nuclear fuel). Thermal breeder reactors previously tested using ^{232}Th to breed the fissile isotope ^{233}U continue to be studied and developed.

While, in principle, all fission reactors can act in all three capacities, in practice the tasks lead to conflicting engineering goals and most reactors have been built with only one of the above tasks in mind. (There are several early counter-examples, such as the Hanford N reactor, now decommissioned).

Power reactors generally convert the kinetic energy of fission products into heat, which is used to heat a working fluid and drive a heat engine that generates mechanical or electrical power. The working fluid is usually water with a steam turbine, but some designs use other materials such as gaseous helium.

Research reactors produce neutrons that are used in various ways, with the heat of fission being treated as an unavoidable waste product. Breeder reactors are a specialized form of research reactor, with the caveat that the sample being irradiated is usually the fuel itself, a mixture of ^{238}U and ^{235}U. For a more detailed description of the physics and operating principles of critical fission reactors, see nuclear reactor physics. For a description of their social, political, and environmental aspects, see nuclear reactor.

Fission Bombs

One class of nuclear weapon, a *fission bomb* (not to be confused with the *fusion bomb*), otherwise known as an *atomic bomb* or *atom bomb*, is a fission reactor designed to liberate as much energy as possible as rapidly as possible, before the released energy causes the reactor to explode (and the chain reaction to stop). Development of nuclear weapons was the motivation behind early research into nuclear fission: the Manhattan Project of the U.S. military during World War II carried out most of the early scientific work on fission chain reactions, culminating in the Trinity test bomb and the Little Boy and Fat Man bombs that were exploded over the cities Hiroshima, and Nagasaki, Japan in August 1945.

Even the first fission bombs were thousands of times more explosive than a comparable mass of chemical explosive. For example, Little Boy weighed a total of about four tons (of which 60 kg was nuclear fuel) and was 11 feet (3.4 m) long; it also yielded an explosion equivalent to about 15 kilotons of TNT, destroying a large part of the city of Hiroshima. Modern nuclear weapons (which include a thermonuclear *fusion* as well as one or more fission stages) are literally hundreds of times more energetic for their weight than the first pure fission atomic bombs, so that a modern single missile warhead bomb weighing less than 1/8th as much as Little Boy has a yield of 475,000 tons of TNT, and could bring destruction to 10 times the city area. While the fundamental physics of the fission chain reaction in a nuclear weapon is similar to the physics of a controlled nuclear reactor, the two types of device must be engineered quite differently.

A nuclear bomb is designed to release all its energy at once, while a reactor is designed to generate a steady supply of useful power. While overheating of a reactor can lead to, and has led to, meltdown and steam explosions, the much lower uranium enrichment makes it impossible for a nuclear reactor to explode with the same destructive power as a nuclear weapon. It is also difficult to extract useful power from a nuclear bomb, although at least one rocket propulsion system, Project Orion, is intended to work by exploding fission bombs behind a massively-padded and shielded vehicle.

The strategic importance of nuclear weapons is a major reason why the technology of nuclear fission is politically sensitive. Viable fission bomb designs are, arguably, within the capabilities of many being relatively simple from an engineering viewpoint. However, the difficulty of obtaining fissile nuclear material to realize the designs, is the key to the relative unavailability of nuclear weapons to all but modern industrialized governments with special programmes to produce fissile materials.

Discovery of Fission

The discovery of nuclear fission occurred in 1938, following nearly five decades of work on the science of radioactivity and the elaboration of new nuclear physics that described the components of atoms. In 1911, Ernest Rutherford proposed a model of the atom in which a very small, dense and positively-charged nucleus of protons was surrounded by orbiting, negatively-charged electrons (the Rutherford model). Niels Bohr improved upon this in 1913 by reconciling the quantum behaviour of electrons (the Bohr model).

Work by Henri Becquerel, Marie Curie, Pierre Curie, and Rutherford further elaborated that the nucleus, though tightly bound, could undergo different forms of radioactive decay, and thereby transmute into other elements (for example, by losing an alpha particle). All known radioactive processes before fission changed mass of the atomic nucleus by no more than two protons and two neutrons (Alpha decay). Albert Einstein's principle of mass–energy equivalence described the amount of energy released in such processes, but this could not be harnessed on a large scale.

The possibility of combining two light nuclei in nuclear fusion had been studied in connection with the processes which power stars. After English physicist James Chadwick discovered the neutron in 1932, Enrico Fermi and his colleagues in Rome studied the results of bombarding uranium with neutrons in 1934. Fermi concluded that his experiments had created a new element with 94 protons, which he dubbed Hesperium. However, not all were convinced with Fermi's analysis of his results.

The German chemist Ida Noddack notably suggested in 1934 that instead of creating a new, heavier element, that "it is conceivable that the nucleus breaks up into several large fragments." However, Noddack's conclusion was not pursued. After the Fermi publication, Lise Meitner, Otto Hahn, and Fritz Strassmann began performing similar experiments in Berlin. Meitner, an Austrian Jew, lost her citizenship with the Anschluss in 1938. She fled and wound up in Sweden, but continued to collaborate by mail and through meetings with Hahn in Sweden.

By coincidence her nephew Otto Robert Frisch, also a refugee, was also in Sweden when Meitner received a letter from Hahn describing his chemical proof that some of the product of the bombardment of uranium with neutrons was barium. Hahn was unsure of what the physical basis for the results were—barium had an atomic mass 40% less than uranium, and no previously known methods of radioactive decay could account for such a radical difference in the size of the nucleus. In Sweden, Frisch was skeptical, but Meitner trusted Hahn's ability as a chemist. Marie Curie had been separating barium from radium for many years, and the techniques were well-known. According to Frisch:

> *Was it a mistake? No, said Lise Meitner; Hahn was too good a chemist for that. But how could barium be formed from uranium? No larger fragments than protons or*

helium nuclei (alpha particles) had ever been chipped away from nuclei, and to chip off a large number not nearly enough energy was available. Nor was it possible that the uranium nucleus could have been cleaved right across. A nucleus was not like a brittle solid that can be cleaved or broken; George Gamow had suggested early on, and Bohr had given good arguments that a nucleus was much more like a liquid drop.

Perhaps a drop could divide itself into two smaller drops in a more gradual manner, by first becoming elongated, then constricted, and finally being torn rather than broken in two? We knew that there were strong forces that would resist such a process, just as the surface tension of an ordinary liquid drop tends to resist its division into two smaller ones. But nuclei differed from ordinary drops in one important way: they were electrically charged, and that was known to counteract the surface tension.

The charge of a uranium nucleus, we found, was indeed large enough to overcome the effect of the surface tension almost completely; so the uranium nucleus might indeed resemble a very wobbly unstable drop, ready to divide itself at the slightest provocation, such as the impact of a single neutron. But there was another problem. After separation, the two drops would be driven apart by their mutual electric repulsion and would acquire high speed and hence a very large energy, about 200 MeV in all; where could that energy come from?...Lise Meitner... worked out that the two nuclei formed by the division of a uranium nucleus together would be lighter than the original uranium nucleus by about one-fifth the mass of a proton. Now whenever mass disappears energy is created, according to Einstein's formula $E=mc^2$*, and one-fifth of a proton mass was just equivalent to 200MeV. So here was the source for that energy; it all fitted!*

In short, Meitner had correctly interpreted Hahn's results to mean that the nucleus of uranium had split roughly in half. Frisch named the process "fission" as an analogy to binary fission in the biological sciences. In December 1938, the German chemists Otto Hahn and Fritz Strassmann sent a manuscript to *Naturwissenschaften* reporting they had detected the element barium after bombarding uranium with neutrons; simultaneously, they communicated these results to Lise Meitner.

Meitner, and her nephew Otto Robert Frisch, correctly interpreted these results as being nuclear fission. Frisch confirmed this experimentally on 13 January 1939. In 1944, Hahn received the Nobel Prize for Chemistry for the discovery of nuclear fission. Some historians who have documented the history of the discovery of nuclear fission believe Meitner should have been awarded the Nobel Prize with Hahn. News spread quickly of the new discovery, which was correctly seen as an entirely novel physical effect with great scientific—and potentially practical—possibilities.

Meitner's and Frisch's interpretation of the work of Hahn and Strassmann crossed the Atlantic Ocean with Niels Bohr, who was to lecture at Princeton University. I.I. Rabi and Willis Lamb, two Columbia University physicists working at Princeton, heard the news and carried it back to Columbia. Rabi said he told Enrico Fermi; Fermi gave credit to Lamb. Bohr soon thereafter went from Princeton to Columbia to see Fermi.

Not finding Fermi in his office, Bohr went down to the cyclotron area and found Herbert L. Anderson. Bohr grabbed him by the shoulder and said: "Young man, let me explain to you about something new and exciting in physics." It was clear to a number of scientists at Columbia that they should try to detect the energy released in the nuclear fission of uranium from neutron bombardment. On 25 January 1939, a Columbia University team conducted the first nuclear fission experiment in the United States, which was done in the basement of Pupin Hall; the members of the team were Herbert L. Anderson, Eugene T. Booth, John R. Dunning, Enrico Fermi, G. Norris Glasoe, and Francis G. Slack.

The experiment involved placing uranium oxide inside of an ionization chamber and irradiating it with neutrons, and measuring the energy thus released. The results confirmed that fission was occurring and hinted strongly that it was the isotope uranium 235 in particular that was fissioning. The next day, the Fifth Washington Conference on Theoretical Physics began in Washington, D.C. under the joint auspices of the George Washington University and the Carnegie Institution of Washington.

There, the news on nuclear fission was spread even further, which fostered many more experimental demonstrations. During this period, the Hungarian physicist Leo Szilαrd, then in the United States, realized that fission could be used to create a nuclear chain reaction, an idea he had first formulated in 1933 but had not been able to achieve with neutron-rich light atoms (nuclear fission then being unknown).

If the number of secondary neutrons produced by each fissioning nucleus was greater than one, then each fission reaction could, in theory, trigger two more reactions. Such a system of exponential growth held out the possibility of using uranium fission as a means to generate large amounts of energy, either for civilian (i.e. electric) purposes, or even for military purposes—an atomic bomb. Szilard urged Fermi (in New York) and Frédéric Joliot-Curie (in Paris) to refrain from publishing on the possibility of a chain reaction, lest the Nazi government become aware of the possibilities on the eve of World War II. Fermi agreed to self-censor, with some hesitation. Joliot-Curie, however, did not, and in April 1939, his team in Paris (Joliot-Curie, Hans von Halban, and Lew Kowarski) reported in *Nature* that the number of neutrons emitted with nuclear fission of ^{235}U was then reported at 3.5 per fission. (They later corrected this to 2.6 per fission.) Simultaneous work by Szilard and Walter Zinn confirmed these results. This appeared to make the possibility of building nuclear reactors, and perhaps nuclear bombs, possible in theory. There was still much unknown about fission and chain reacting systems, however.

The Fission Chain Reaction

"Chain reactions" at that time were a known phenomenon in *chemistry*, but the analogous process in nuclear physics, using neutrons, had been foreseen as early as 1933 by Szilαrd, although Szilαrd at that time had no idea with what materials the process might be initiated. Szilαrd considered that neutrons would be ideal for such a situation, since they lacked an electrostatic charge. With the news of fission neutrons from uranium fission, Szilαrd immediately understood the possibility of a nuclear chain reaction using uranium. In the summer, Fermi and Szilard proposed the idea of a nuclear reactor (pile) to mediate this process.

The pile would use natural uranium as fuel. Fermi had shown much earlier that neutrons were far more effectively captured by atoms if they were of low energy (so-called "slow" or "thermal" neutrons), because for quantum reasons it made the atoms look like much larger targets to the neutrons. Thus to slow down the secondary neutrons released by the fissioning uranium nuclei, Fermi and Szilard proposed a graphite "moderator," against which the fast, high-energy secondary neutrons would collide, effectively slowing them down. With enough uranium, and with pure-enough graphite, their "pile" could theoretically sustain a slow-neutron chain reaction. This would result in the production of heat, as well as the creation of radioactive fission products. In August 1939, Szilard and fellow Hungarian refugees physicists Teller and Wigner thought that the Germans might make use of the fission chain reaction and were spurred to attempt to attract the attention of the United States government to the issue.

Towards this, they persuaded German-Jewish refugee Albert Einstein to lend his name to a letter directed to President Franklin Roosevelt. The Einstein–Szilard letter suggested the possibility of a uranium bomb deliverable by ship, which would destroy "an entire harbor and much of the surrounding countryside." The President received the letter on 11 October 1939 — shortly after World War II began in Europe, but two years before U.S. entry into it. Roosevelt ordered that a scientific committee be authorized for overseeing uranium work and allocated a small sum of money for pile research.

In England, James Chadwick proposed an atomic bomb utilizing natural uranium, based on a paper by Rudolf Peierls with the mass needed for critical state being 30–40 tons. In America, J. Robert Oppenheimer thought that a cube of uranium deuteride 10 cm on a side (about 11 kg of uranium) might "blow itself to hell." In this design it was still thought that a moderator would need to be used for nuclear bomb fission (this turned out not to be the case if the fissile isotope was separated). In December, Heisenberg delivered a report to the German Ministry of War on the possibility of a uranium bomb.

Most of these models were still under the assumption that the bombs would be powered by slow neutron reactions—and thus be similar to a reactor undergoing a meltdown. In Birmingham, England, Frisch teamed up with Peierls, a fellow German-Jewish refugee. They had the idea of using a purified mass of the uranium isotope ^{235}U, which had a cross section just determined, and which was much larger than that of ^{238}U or natural uranium (which is 99.3% the latter isotope). Assuming that the cross section for fast-neutron fission of ^{235}U was the same as for slow neutron fission, they determined that a pure ^{235}U bomb could have a critical mass of only 6 kg instead of tons, and that the resulting explosion would be tremendous. (The amount actually turned out to be 15 kg, although several times this amount was used in the actual uranium (Little Boy) bomb). In February 1940 they delivered the Frisch–Peierls memorandum. Ironically, they were still officially considered "enemy aliens" at the time. Glenn Seaborg, Joseph W. Kennedy, Arthur Wahl and Italian-Jewish refugee Emilio Segrè shortly discovered ^{239}Pu in the decay products of ^{239}U produced by bombarding ^{238}U with neutrons, and determined it to be a fissile material, like ^{235}U.

The possibility of isolating uranium-235 was technically daunting, because uranium-235 and uranium-238 are chemically identical, and vary in their mass by only the weight of three neutrons. However, if a sufficient quantity of uranium-235 could be isolated, it would allow for a fast neutron fission chain reaction. This would be extremely explosive, a true "atomic bomb." The discovery that plutonium-239 could be produced in a nuclear reactor pointed towards another approach to a fast neutron fission bomb.

Both approaches were extremely novel and not yet well understood, and there was considerable scientific skepticism at the idea that they could be developed in a short amount of time. On June 28, 1941, the Office of Scientific Research and Development was formed in the U.S. to mobilize scientific resources and apply the results of research to national defense. In September, Fermi assembled his first nuclear "pile" or reactor, in an attempt to create a slow neutron-induced chain reaction in uranium, but the experiment failed to achieve criticality, due to lack of proper materials, or not enough of the proper materials which were available.

Producing a fission chain reaction in natural uranium fuel was found to be far from trivial. Early nuclear reactors did not use isotopically enriched uranium, and in consequence they were required to use large quantities of highly purified graphite as neutron moderation materials. Use of ordinary water (as opposed to heavy water) in nuclear reactors requires enriched fuel — the partial separation and relative enrichment of the rare ^{235}U isotope from the far more common ^{238}U isotope. Typically, reactors also require inclusion of extremely chemically pure neutron moderator materials such as deuterium (in heavy water), helium, beryllium, or carbon, the latter usually as graphite. (The high purity for carbon is required because many chemical impurities such as the boron-10 component of natural boron, are very strong neutron absorbers and thus poison the chain reaction and ending it prematurely.)

Production of such materials at industrial scale had to be solved for nuclear power generation and weapons production to be accomplished. Up to 1940, the total amount of uranium metal produced in the USA was not more than a few grams, and even this was of doubtful purity; of metallic beryllium not more than a few kilograms; and concentrated deuterium oxide (heavy water) not more than a few kilograms. Finally, carbon had never been produced in quantity with anything like the purity required of a moderator.

The problem of producing large amounts of high purity uranium was solved by Frank Spedding using the thermite or "Ames" process. Ames Laboratory was established in 1942 to produce the large amounts of natural (unenriched) uranium metal that would be necessary for the research to come. The critical nuclear chain-reaction success of the Chicago Pile-1 (December 2, 1942) which used unenriched (natural) uranium, like all of the atomic "piles" which produced the plutonium for the atomic bomb, was also due specifically to Szilard's realization that very pure graphite could be used for the moderator of even natural uranium "piles". In wartime Germany, failure to appreciate the qualities of very pure graphite led to reactor designs dependent on heavy water, which in turn was denied the Germans by Allied attacks in Norway, where heavy water was produced. These difficulties—among many others— prevented the Nazis from building a nuclear reactor capable of criticality during the war.

Manhattan Project and Beyond

In the United States, an all-out effort for making atomic weapons was begun in late 1942. This work was taken over by the U.S. Army Corps of Engineers in 1943, and known as the Manhattan Engineer District. The top-secret Manhattan Project, as it was colloquially known, was led by General Leslie R. Groves. Among the project's dozens of sites were: Hanford Site in Washington state, which had the first industrial-scale nuclear reactors; Oak Ridge, Tennessee, which was primarily concerned with uranium enrichment; and Los Alamos, in New Mexico, which was the scientific hub for research on bomb development and design. Other sites, notably the Berkeley Radiation Laboratory and the Metallurgical Laboratory at the University of Chicago, played important contributing roles.

Overall scientific direction of the project was managed by the physicist J. Robert Oppenheimer. In July 1945, the first atomic bomb, dubbed "Trinity", was detonated in the New Mexico desert. It was fueled by plutonium created at Hanford. In August 1945, two more atomic bombs—"Little Boy", a uranium-235 bomb, and "Fat Man", a plutonium bomb—were used against the Japanese cities of Hiroshima and Nagasaki. In the years after World War II, many countries were involved in the further development of nuclear fission for the purposes of nuclear reactors and nuclear weapons.

Natural Fission Chain-Reactors on Earth

Criticality in nature is uncommon. At three ore deposits at Oklo in Gabon, sixteen sites (the so-called Oklo Fossil Reactors) have been discovered at which self-sustaining nuclear fission took place approximately 2 billion years ago. Unknown until 1972 (but postulated by Paul Kuroda in 1956), when French physicist Francis Perrin discovered the Oklo Fossil Reactors, it was realized that nature had beaten humans to the punch.

Large-scale natural uranium fission chain reactions, moderated by normal water, had occurred far in the past and would not be possible now. This ancient process was able to use normal water as a moderator only because 2 billion years before the present, natural uranium was richer in the shorter-lived fissile isotope ^{235}U (about 3%), than natural uranium available today (which is only 0.7%, and must be enriched to 3% to be usable in light-water reactors).

Chapter 7

Atomic Structure of Elements

Atomic Nucleus

The nucleus is the very dense region consisting of nucleons (protons and neutrons) at the center of an atom. Almost all of the mass in an atom is made up from the protons and neutrons in the nucleus, with a very small contribution from the orbiting electrons. It was discovered in 1911, as a result of Ernest Rutherford's interpretation of the famous 1909 Rutherford experiment performed by Hans Geiger and Ernest Marsden, under the direction of Rutherford. The proton-neutron model of nucleus was proposed by Dmitry Ivanenko in 1932. The diameter of the nucleus is in the range of 1.75 fm (1.75×10^{-15} m) for hydrogen (the diameter of a single proton) to about 15 fm for the heaviest atoms, such as uranium. These dimensions are much smaller than the diameter of the atom itself (nucleus + electronic cloud), by a factor of about 23,000 (uranium) to about 145,000 (hydrogen). The branch of physics concerned with studying and understanding the atomic nucleus, including its composition and the forces which bind it together, is called nuclear physics.

Etymology

The term nucleus is from the Latin word *nucleus* ("nut"). In 1844, Michael Faraday used the term to refer to the "central point of an atom". The modern atomic meaning was proposed by Ernest Rutherford in 1912. The adoption of the term "nucleus" to atomic theory, however, was not immediate. In 1916, for example, Gilbert N. Lewis stated, in his famous article *The Atom and the Molecule,* that "the atom is composed of the *kernel* and an outer atom or *shell*"

Nuclear Makeup

The nucleus of an atom consists of protons and neutrons (two types of baryons) bound by the nuclear force (also known as the residual strong force). These baryons are further composed of subatomic fundamental particles known as quarks bond by the strong interaction. Which chemical element an atom represents is determined by the number of protons in the nucleus. Each proton carries a single positive charge, and the total electrical charge of the nucleus is spread fairly uniformly throughout its body, with a fall-off at the edge. Major exceptions to this rule are the light elements hydrogen and helium, where the charge is concentrated most highly at the single central point (without a volume of uniform charge), as would be expected for fermions (in this case, protons) in 1s states without orbital angular momentum. As each proton carries a unit of charge, the charge distribution is indicative of the proton distribution. The neutron distribution probably is similar.

Protons and Neutrons

Protons and neutrons are fermions, with different values of the isospin quantum number, so two protons and two neutrons can share the same space wave function since they are not identical quantum entities. They sometimes are viewed as two different quantum states of the same particle, the *nucleon*. Two fermions, such as two protons, or two neutrons, or a proton + neutron (the deuteron) can exhibit bosonic behaviour when they become loosely bound in pairs. In the rare case of a hypernucleus, a third baryon called a hyperon, with a different value of the strangeness quantum number can also share the wave function. However, the latter type of nuclei are extremely unstable and are not found on Earth except in high energy physics experiments. The neutron has a positively charged core of radius H" 0.3 fm surrounded by a compensating negative charge of radius between 0.3 fm and 2 fm. The proton has an approximately exponentially decaying positive charge distribution with a mean square radius of about 0.8 fm.

Forces

Nuclei are bound together by the residual strong force (nuclear force). The residual strong force is minor residuum of the strong interaction which binds quarks together to form protons and neutrons. This force is much weaker *between* neutrons and protons because it is mostly neutralized within them, in the same way that electromagnetic forces *between* neutral atoms (such as van der Waals forces that act between two inert gas atoms) are much weaker than the electromagnetic forces that hold the parts of the atoms internally together (for example, the forces that hold the electrons in an inert gas atom bound to its nucleus).

The nuclear force is highly attractive at the distance of typical nucleon separation, and this overwhelms the repulsion between protons which is due to the electromagnetic force, thus allowing nuclei to exist. However, because the residual strong force has a limited range because it decays quickly with distance, only nuclei smaller than a certain size can be completely stable. The largest known completely stable (e.g., stable to alpha, beta, and gamma decay) nucleus is lead-208 which contains a total of 208 nucleons (126 neutrons and 82 protons). Nuclei larger than this maximal size of 208 particles are unstable and (as a trend) become increasingly short-lived with larger size, as the number of neutrons and protons which compose them increases beyond this number. However, bismuth-209 is also stable to beta decay and has the longest half-life to alpha decay of any known isotope, estimated at a billion times longer than the age of the universe. The residual strong force is effective over a very short range (usually only a few fermis; roughly one or two nucleon diameters) and causes an attraction between any pair of nucleons. For example, between protons and neutrons to form [NP] deuteron, and also between protons and protons, and neutrons and neutrons. It also is effective for the stability of one 3-body nucleon system [PNP], helium-3, while the triton [NPN] is unstable and decays to helium-3.

Halo Nuclei and Strong Force Range Limits

The effective absolute limit of the range of the strong force is represented by halo nuclei such as lithium-11 or boron-14, in which dineutrons, or other collections of neutrons, orbit at distances of about ten fermis (roughly similar to the 8 fermi radius of the nucleus of uranium-238). These nuclei are not maximally dense. Halo nuclei form at the extreme edges of the chart of the nuclides—the neutron drip line and proton drip line—and are all unstable with short half-lives, measured in milliseconds; for example, lithium-11 has a half-life of less than 8.6 milliseconds. Halos in effect represent an excited state with nucleons in an outer quantum shell which has unfilled energy levels "below" it (both in terms of radius and energy). The halo may be made of either neutrons [NN, NNN] or protons [PP, PPP]. Examples: Nuclei which have a single neutron halo include 11Be and 19C. A two-neutron halo is exhibited by ^{6}He, ^{11}Li, ^{17}B, ^{19}B and ^{22}C. Two-neutron halo nuclei break into three fragments, never two, and are called *Borromean* because of this behaviour (referring to a system of three interlocked rings in which breaking any ring frees both of the others). ^{8}He and ^{14}Be both exhibit a four-neutron halo. Nuclei which have a proton halo include ^{8}B and ^{26}P. A two-proton halo is exhibited by ^{17}Ne and ^{27}S. Proton halos are expected to be more rare and unstable than the neutron examples, because of the repulsive electromagnetic forces of the excess proton(s).

Nuclear Models

There are many different historical models of the atomic nucleus, none of which to this day completely explains experimental data on nuclear structure. A useful review of 37 known models of the atomic nucleus is provided by Cook. In other words, packing protons and neutrons in the nucleus gives *approximately* the same total size result as packing hard spheres of a constant size (like marbles) into a tight spherical or semi-spherical bag (some stable nuclei are not quite spherical, but are known to be prolate).

Liquid Drop Models

Early models of the nucleus viewed the nucleus as a rotating liquid drop. In this model, the trade-off of long-range electromagnetic forces and relatively short-range nuclear forces, together cause behaviour which resembled surface tension forces in liquid drops of different sizes. This formula is successful at explaining many important phenomena of nuclei, such as their changing amounts of binding energy as their size and composition changes, but it does not explain the special stability which occurs when nuclei have special "magic numbers" of protons or neutrons.

Shell Models and Other Quantum Models

A number of models for the nucleus have also been proposed in which nucleons occupy orbitals, much like the atomic orbitals in atomic physics theory. These wave models imagine nucleons to be either sizeless point particles in potential wells, or else probability waves as in the "optical model", frictionlessly orbiting at high speed in potential wells. In these models, the nucleons may occupy orbitals in pairs, due to being fermions, but the exact nature and capacity of nuclear shells differs from those of electrons in atomic orbitals, primarily because the potential well in which the nucleons move (especially in larger nuclei) is quite different from the central electromagnetic potential well which binds electrons in atoms. Some resemblance to atomic orbital models may be seen in a small atomic nucleus like that of helium-4, in which the two protons and two neutrons separately occupy 1s orbitals analogous to the 1s orbital for the two electrons in the helium atom, and achieve unusual stability for the same reason. Nuclei with 5 nucleons are all extremely unstable and short-lived, yet, helium-3, with 3 nucleons, is very stable even with lack of a closed 1s orbital shell. Another nucleus with 3 nucleons, the triton hydrogen-3 is unstable and will decay into helium-3 when isolated. Weak nuclear stability with 2 nucleons {NP} in the 1s orbital is found in the deuteron hydrogen-2, with only one nucleon in each of the proton and neutron potential wells. While each nucleon is a fermion, the {NP} deuteron is a boson and thus does not follow Pauli Exclusion for close packing within shells. Lithium-6 with 6 nucleons is highly stable without a closed second 1p shell orbital. For light nuclei with total nucleon numbers 1 to 6 only those with 5 do not show some evidence of stability.

Observations of beta-stability of light nuclei outside closed shells indicate that nuclear stability is much more complex than simple closure of shell orbitals with magic numbers of protons and neutrons. For larger nuclei, the shells occupied by nucleons begin to differ significantly from electron shells, but nevertheless, present nuclear theory does predict the magic numbers of filled nuclear shells for both protons and neutrons.

The closure of the stable shells predicts unusually stable configurations, analogous to the noble group of nearly-inert gases in chemistry. An example is the stability of the closed shell of 50 protons, which allows tin to have 10 stable isotopes, more than any other element. Similarly, the distance from shell-closure explains the unusual instability of isotopes which have far from stable numbers of these particles, such as the radioactive elements 43 (technetium) and 61 (promethium), each of which is preceded and followed by 17 or more stable elements. There are however problems with the shell model when an attempt is made to account for nuclear properties well away from closed shells. This has led to complex *post hoc* distortions of the shape of the potential well to fit experimental data, but the question remains whether these mathematical manipulations actually correspond to the spatial deformations in real nuclei. Problems with the shell model have led some to propose realistic two-body and three-body nuclear force effects involving nucleon clusters and then build the nucleus on this basis. Two such cluster models are the Close-Packed Spheron Model of Linus Pauling and the 2D Ising Model of MacGregor.

Consistency between Models

As with the case of superfluid liquid helium, atomic nuclei are an example of a state in which both (1) "ordinary" particle physical rules for volume and (2) non-intuitive quantum mechanical rules for a wave-like nature apply. In superfluid helium, the helium atoms have volume, and essentially "touch" each other, yet at the same time exhibit strange bulk properties, consistent with a Bose-Einstein condensation. The latter reveals that they also have a wave-like nature and do not exhibit standard fluid properties, such as friction. For nuclei made of hadrons which are fermions, the same type of condensation does not occur, yet nevertheless, many nuclear properties can only be explained similarly by a combination of properties of particles with volume, in addition to the frictionless motion characteristic of the wave-like behaviour of objects trapped in Schrödinger quantum orbitals.

Background Radiation

Background radiation is radiation that is constantly present in the environment and is emitted from a variety of natural and artificial sources. Primary contributions come from:

- Sources in the earth. These include sources in food and water, which are incorporated in the body, and in building materials and other products that incorporate those radioactive sources;
- Sources from space, in the form of cosmic rays;
- Sources in the atmosphere. One significant contribution comes from the radon gas that is released from the Earth's crust and subsequently decays into radioactive atoms that become attached to airborne dust and particulates. Another contribution arises from the radioactive atoms produced in the bombardment of atoms in the upper atmosphere by high-energy cosmic rays.

Naturally occurring sources are responsible for the vast majority of radiation exposure. However, not including direct exposure from radiological imaging or therapy, about 3% of background radiation comes from man-made sources such as:

- Self-luminous dials and signs
- Global radioactive contamination due to historical nuclear weapons testing
- Nuclear power station or nuclear fuel reprocessing accidents
- Normal operation of facilities used for nuclear power and scientific research
- Emissions from burning fossil fuels, such as coal fired power plants
- Emissions from nuclear medicine facilities and patients

Natural Background Radiation

Natural background radiation comes from two primary sources: cosmic radiation and terrestrial sources. The worldwide average background dose for a human being is about 2.4 millisievert (mSv) per year. This exposure is mostly from cosmic radiation and natural radionuclides in the environment (including those within the body). This is far greater than human-caused background radiation exposure, which in the year 2000 amounted to an average of about 5 ìSv per year from historical nuclear weapons testing, nuclear power accidents and nuclear industry operation combined, and is greater than the average exposure from medical tests, which ranges from 0.04 to 1 mSv per year. Older coal-fired power plants without effective fly ash capture are one of the largest sources of human-caused background radiation exposure. The level of natural background radiation varies depending on location, and in some areas the level is significantly higher than average. Such areas include Ramsar in Iran, Guarapari in Brazil, Kerala in India, the northern Flinders Ranges in Australia and Yangjiang in China. In Ramsar a peak yearly dose of 260 mSv has been reported (compared with 0.06 mSv of a chest radiograph or up to 20 mSv of a CT scan).

The highest levels of natural background radiation recorded in the world is from areas around Ramsar, particularly at Talesh-Mahalleh which is a very high background radiation area (VHBRA) having an effective dose equivalent several times in excess of ICRP-recommended radiation dose limits for radiation workers and up to 200 times greater than normal background levels. Most of the radiation in the area is due to dissolved radium-226 in water of hot springs along with smaller amounts of uranium and thorium due to travertine deposits. There are more than nine hot springs in the area with different concentrations of radioisotopes, and these are used as spas by locals and tourists. This high level of radiation does not seem to have caused ill effects on the residents of the area and even possibly has made them slightly more radioresistant, which is puzzling and has been called "radiation paradox". It has also been reported that residents have healthier and longer lives. On the basis of this and other evidences including the fact that life had originated in a much more irradiated environment, some scientists have questioned the validity of linear no-threshold model, on which all radiation regulations currently depend. Others point out that some level of radiation might actually be good for health and have a positive effect on population based on the controversial radiation hormesis model, by jump starting DNA repair mechanisms inside the cell. Background radiation doses in the immediate vicinities of particles of high atomic number materials, within the human body, have a small enhancement due to the photoelectric effect.

Cosmic Radiation

The Earth, and all living things on it, are constantly bombarded by radiation from outer space. This radiation primarily consists of positively charged ions from protons to iron and larger nuclei derived sources outside our solar system. This radiation interacts with atoms in the atmosphere to create secondary radiation, including X-rays, muons, protons, alpha particles, pions, electrons, and neutrons. The immediate dose from cosmic radiation is largely from muons, neutrons, and electrons, and this dose varies in different parts of the world based largely on the geomagnetic field and altitude. This radiation is much more intense in the upper troposphere, around 10 km altitude, and is thus of particular concern for airline crews and frequent passengers, who spend many hours per year in this environment. During their flights airline crews typically get an extra dose on the order of 2.2mSv (220 mrem) per year. Similarly, cosmic rays cause higher background exposure in astronauts than in humans on the surface of Earth. Astronauts in low orbits, such as in the International Space Station or the Space Shuttle, are partially shielded by the magnetic field of the Earth, but also suffer from the Van Allen radiation belt which accumulates cosmic rays and results from the earths magnetic field. Outside low Earth orbit, as experienced by the Apollo astronauts who traveled to the Moon, this background radiation is much more intense, and represents a considerable obstacle to potential future long term human exploration of the moon or Mars.

Cosmic rays also cause elemental transmutation in the atmosphere, in which secondary radiation generated by the cosmic rays combines with atomic nuclei in the atmosphere to generate different nuclides. Many so-called cosmogenic nuclides can be produced, but probably the most notable is carbon-14, which is produced by interactions with nitrogen atoms. These cosmogenic nuclides eventually reach the Earth's surface and can be incorporated into living organisms. The production of these nuclides varies slightly with short-term variations in solar cosmic ray flux, but is considered practically constant over long scales of thousands to millions of years. The constant production, incorporation into organisms and relatively short half-life of carbon-14 are the principles used in radiocarbon dating of ancient biological materials such as wooden artifacts or human remains.

Terrestrial Sources

Radioactive material is found throughout nature. It occurs naturally in the soil, rocks, water, air, and vegetation. The major radionuclides of concern for terrestrial radiation are common elements with low-abundance radioactive isotopes, like potassium and carbon, or the long-lived elements uranium and thorium and their decay products, some of which, like radium and radon are intensely radioactive but occur in low concentration. Most of these sources have been decreasing, due to radioactive decay since the formation of the Earth, because there is no significant amount currently transported to the Earth. Thus, the present activity on earth from uranium-238 is only half as much as it originally was because of its 4.5 billion year half-life, and potassium-40 (half life 1.25 billion years) is only at about 8% of original activity.

The effects on humas of the actual diminishment (due to decay) of these isotopes is minimal however. This is because humans evolved too recently for the difference in activity over a fraction of a half-life to be significant. Put another way, human history is so short in comparison to a half life of a billion years, that the activity of these long-lived isotopes has been effectively constant throughout our time on this planet. In addition, many shorter half-life and thus more intensely radioactive isotopes have not decayed out of the terrestrial environment, however, because of natural on-going production of them. Examples of these are carbon-14 (cosmogenic), radium-226 (decay product of uranium-238) and radon-222 (a decay product of radium-226).

Radiation Inside the Human Body

Some of the essential elements that make up the human body, mainly potassium and carbon, have radioactive isotopes that add significantly to our background radiation dose. An average human contains about 30 milligrams of potassium-40 (^{40}K) and about 10 nanograms (10^{-8} g) of carbon-14 (^{14}C). Excluding internal contamination by external radioactive material, the largest component of internal radiation exposure from biologically functional components of the human body is from potassium-40.

The decay of about 4,000 nuclei of ^{40}K per second makes potassium the largest source of radiation in terms of number of decaying atoms. The energy of beta particles produced by ^{40}K is also about 10 times more powerful than the beta particles from ^{14}C decay. There are about 1,200 beta particles per second produced by the decay of ^{14}C. However, a ^{14}C atom is in the genetic information of about half the cells, while potassium is not a component of DNA. The decay of a ^{14}C atom inside DNA in one person happens about 50 times per second, changing a carbon atom to one of nitrogen.

Radon

Radon is a terrestrial source of ionizing radiation that is of particular concern because, although on average it is very rare, this intensely radioactive element can be found in high concentrations in many areas of the world, where it represents a significant health hazard. Radon is a decay product of uranium, which is relatively common in the Earth's crust, but generally concentrated in ore-bearing rocks scattered around the world.

Radon seeps out of these ores into the atmosphere or into ground water, and in these localities it can accumulate within dwellings and expose humans to high concentrations. The widespread construction of well insulated and sealed homes in the northern industrialized world has led to radon becoming the primary source of background radiation in some localities in northern North America and Europe. Some of these areas, including Cornwall and Aberdeenshire in the United Kingdom have high enough natural radiation levels that nuclear licensed sites cannot be built there — the sites would already exceed legal radiation limits before they opened, and the natural topsoil and rock would all have to be disposed of as low-level nuclear waste. Radiation exposure from radon is indirect. Radon has a short half-life (4 days) and decays into other solid particulate radium-series radioactive nuclides. These radioactive particles are inhaled and remain lodged in the lungs, causing continued exposure. People in affected localities can receive up to 10 mSv per year background radiation. Radon is thus the second leading cause of lung cancer after smoking, and accounts for 15,000 to 22,000 cancer deaths per year in the US alone.

Human-caused Background Radiation

Frequent above-ground nuclear explosions between the 1940s and 1960s scattered a substantial amount of radioactive contamination. Some of this contamination is local, rendering the immediate surroundings highly radioactive, while some of it is carried longer distances as nuclear fallout; some of this material is dispersed worldwide. The increase in background radiation due to these tests peaked in 1963 at about 0.15 mSv per year worldwide, or about 7% of average background dose from all sources.

The Limited Test Ban Treaty of 1963 prohibited above-ground tests, thus by the year 2000 the worldwide dose from these historical tests has decreased to only 0.005 mSv per year. Older coal-fired power plants without effective fly ash capture are one of the largest sources of human-caused background radiation exposure. When coal is burned, uranium, thorium and all the uranium daughters accumulated by disintegration — radium, radon, polonium — are released. According to a 1978 article in Science magazine, "coal-fired power plants throughout the world are the major sources of radioactive materials released to the environment". Radioactive materials previously buried underground in coal deposits are released as fly ash or, if fly ash is captured, may be incorporated into concrete manufactured with fly ash. Radioactive materials are also released in gaseous emissions.

The United Nations Scientific Committee on the Effects of Atomic Radiation estimates that per gigawatt-year ($GW_e a$) of electrical energy produced by coal, using the current mix of technology throughout the world, the population impact is approximately 0.8 lethal cancers per plant-year distributed over the affected population.

With 400 GW of coal-fired power plants in the world, this amounts to some 320 deaths per year. Nuclear reactors may also release a certain amount of radioactive contamination. Under normal circumstances, a modern nuclear reactor releases minuscule amounts of radioactive contamination. Major accidents, which have been rare, have also released some radioactive contamination into the environment; this was the case with the Windscale fire (Sellafield accident), the Chernobyl accident, and the Fukushima I nuclear accidents. The amount of radioactive contamination released by human activity is rather small and in most cases negligible in comparison to natural background radiation.

Artificial Radiation Sources

The radiation from natural and artificial radiation sources are identical in their nature and their effects. These materials are distributed in the environment, and in our bodies, according to the chemical properties of the elements. The Nuclear Regulatory Commission, the United States Environmental Protection Agency, and other U.S. and international agencies, require that licensees limit radiation exposure to individual members of the public to 1 mSv (100 mrem) per year, and limit occupational radiation exposure to adults working with radioactive material to 50 mSv (5 rem) per year, and 100 mSv (10 rem) in 5 years. The exposure for an average person is about 3.6 mSv/year, 80 percent of which comes from natural sources of radiation. The remaining 20 percent results from exposure to artificial radiation sources, such as medical x-rays, industrial sources like smoke detectors and a small fraction from nuclear weapons tests. For average persons who have had no medical x-rays, only 3% of their annual radiation dose comes from artificial sources. A standard medical x-ray's strength is about 2 mrem or 0.02 mSv but can be over ten times that, depending on the equipment used. A dental x-ray optimally has a dose as low as 0.0033 mSv but poor machines and technique can give doses as high as 0.11 mSv. The average American and European receives about 0.5 mSv of diagnostic medical dose per year; countries with lower levels of health care receive about one fifth of this dose. Radiation treatment for various diseases also accounts for some dose, both in individuals and in those around them.

Other Usage

In other contexts, background radiation may simply be any radiation that is pervasive, whether ionizing or not. A particular example of this is the cosmic microwave background radiation, a nearly uniform glow that fills the sky in the microwave part of the spectrum; stars, galaxies and other objects of interest in radio astronomy stand out against this background. In a laboratory, background radiation refers to the measured value from any sources that affect an instrument when a radiation source sample is not being measured. This background rate, which must be established as a stable value by multiple measurements, usually before and after sample measurement, is subtracted from the rate measured when the sample is being measured. Background radiation for occupational doses measured for workers is all radiation dose that is not measured by radiation dose measurement instruments in potential occupational exposure conditions. This includes both "natural background radiation" and any medical radiation doses. This value is not typically measured or known from surveys, such that variations in the total dose to individual workers is not known. This can be a significant confounding factor in assessing radiation exposure effects in a population of workers who may have significantly different natural background and medical radiation doses. This is most significant when the occupational doses are very low.

Magnetic Resonance Imaging

Magnetic resonance imaging (MRI), nuclear magnetic resonance imaging (NMRI), or magnetic resonance tomography (MRT) is a medical imaging technique used in radiology to visualize detailed internal structures. MRI makes use of the property of Nuclear magnetic resonance (NMR) to image nuclei of atoms inside the body. An MRI machine uses a powerful magnetic field to align the magnetization of some atoms in the body, and radio frequency fields to systematically alter the alignment of this magnetization. This causes the nuclei to produce a rotating magnetic field detectable by the scanner—and this information is recorded to construct an image of the scanned area of the body. Strong magnetic field gradients cause nuclei at different locations to rotate at different speeds. 3-D spatial information can be obtained by providing gradients in each direction. MRI provides good contrast between the different soft tissues of the body, which make it especially useful in imaging the brain, muscles, the heart, and cancers compared with other medical imaging techniques such as computed tomography (CT) or X-rays. Unlike CT scans or traditional X-rays, MRI uses no ionizing radiation.

How MRI Works

The body is largely composed of water molecules. Each water molecule has two hydrogen nuclei or protons. When a person goes inside the powerful magnetic field of the scanner, the magnetic moments of some of these protons change and align with the direction of the field. In an MRI machine a radio frequency transmitter is briefly turned on, producing an electromagnetic field. The photons of this field have just the right energy, known as the resonance frequency, to flip the spin of the aligned protons in the body. As the intensity and duration of application of the field increase, more aligned spins are affected. After the field is turned off, the protons decay to the original spin-down state and the difference in energy between the two states is released as a photon. It is these photons that produce the electromagnetic signal that the scanner detects.

The frequency at which the protons resonate depends on the strength of the magnetic field. As a result of conservation of energy, this resonation frequency also dictates the frequency of the released photons. The photons released when the field is removed have an energy — and therefore a frequency — due to the amount of energy the protons absorbed while the field was active. It is this relationship between field-strength and frequency that allows the use of nuclear magnetic resonance for imaging. Additional magnetic fields are applied during the scan to make the magnetic field strength depend on the position within the patient, in turn making the frequency of the released photons dependent on position in a predictable manner. Position information can then be recovered from the resulting signal by the use of a Fourier transform.

These fields are created by passing electric currents through specially-wound solenoids, known as gradient coils. Since these coils are within the bore of the scanner, there are large forces between them and the main field coils, producing most of the noise that is heard during operation. Without efforts to damp this noise, it can approach 130 decibels (dB) with strong fields . An image can be constructed because the protons in different tissues return to their equilibrium state at different rates, which is a difference that can be detected. Five different tissue variables — spin density, T_1 and T_2 relaxation times and flow and spectral shifts can be used to construct images. By changing the parameters on the scanner, this effect is used to create contrast between different types of body tissue or between other properties, as in fMRI and diffusion MRI. Contrast agents may be injected intravenously to enhance the appearance of blood vessels, tumors or inflammation. Contrast agents may also be directly injected into a joint in the case of arthrograms, MRI images of joints. Unlike CT, MRI uses no ionizing radiation and is generally a very safe procedure. Nonetheless the strong magnetic fields and radio pulses can affect metal implants, including cochlear implants and cardiac pacemakers. In the case of cochlear implants, the US FDA has approved some implants for MRI compatibility. In the case of cardiac pacemakers, the results can sometimes be lethal, so patients with such implants are generally not eligible for MRI. MRI is used to image every part of the body, and is particularly useful for tissues with many hydrogen nuclei and little density contrast, such as the brain, muscle, connective tissue and most tumors.

History

Magnetic resonance imaging is a relatively new technology. The first MR image was published in 1973 and the first cross-sectional image of a living mouse was published in January 1974. The first studies performed on humans were published in 1977. By comparison, the first human X-ray image was taken in 1895.

Applications

In clinical practice, MRI is used to distinguish pathologic tissue (such as a brain tumor) from normal tissue. One advantage of an MRI scan is that it is harmless to the patient. It uses strong magnetic fields and non-ionizing radiation in the radio frequency range, unlike CT scans and traditional X-rays, which both use ionizing radiation. While CT provides good spatial resolution (the ability to distinguish two separate structures an arbitrarily small distance from each other), MRI provides comparable resolution with far better contrast resolution (the ability to distinguish the differences between two arbitrarily similar but not identical tissues).

The basis of this ability is the complex library of *pulse sequences* that the modern medical MRI scanner includes, each of which is optimized to provide *image contrast* based on the chemical sensitivity of MRI. For example, with particular values of the *echo time* (T_E) and the *repetition time* (T_R), which are basic parameters of image acquisition, a sequence takes on the property of T_2-weighting. On a T_2-weighted scan, water- and fluid-containing tissues are bright (most modern T_2 sequences are actually *fast* T_2 sequences) and fat-containing tissues are dark.

The reverse is true for T_1-weighted images. Damaged tissue tends to develop edema, which makes a T_2-weighted sequence sensitive for pathology, and generally able to distinguish pathologic tissue from normal tissue. With the addition of an additional radio frequency pulse and additional manipulation of the magnetic gradients, a T_2-weighted sequence can be converted to a FLAIR sequence, in which free water is now dark, but edematous tissues remain bright.

This sequence in particular is currently the most sensitive way to evaluate the brain for demyelinating diseases, such as multiple sclerosis. The typical MRI examination consists of 5–20 sequences, each of which are chosen to provide a particular type of information about the subject tissues. This information is then synthesized by the interpreting physician.

Basic MRI Scans

T_1-*Weighted MRI:* T_1-weighted scans are a standard basic scan, in particular differentiating fat from water - with water darker and fat brighter use a gradient echo (GRE) sequence, with short T_E and short T_R. This is one of the basic types of MR contrast and is a commonly run clinical scan.

The T_1 weighting can be increased (improving contrast) with the use of an inversion pulse as in an MP-RAGE sequence. Due to the short repetition time (T_R) this scan can be run very fast allowing the collection of high resolution 3D datasets. A T_1 reducing gadolinium contrast agent is also commonly used, with a T_1 scan being collected before and after administration of contrast agent to compare the difference. In the brain T_1-weighted scans provide good gray matter/white matter contrast; in other words, T_1-weighted images highlight fat deposition.

Spin Density Weighted MRI

Spin density, also called proton density, weighted scans try to have no contrast from either T_2 or T_1 decay, the only signal change coming from differences in the amount of available spins (hydrogen nuclei in water). It uses a spin echo or sometimes a gradient echo sequence, with short T_E and long T_R.

Specialized MRI Scans

Diffusion MRI: Diffusion MRI measures the diffusion of water molecules in biological tissues. In an isotropic medium (inside a glass of water for example) water molecules naturally move randomly according to turbulence and Brownian motion. In biological tissues however, where the Reynold's number is low enough for flows to be laminar, the diffusion may be anisotropic. For example a molecule inside the axon of a neuron has a low probability of crossing the myelin membrane. Therefore the molecule moves principally along the axis of the neural fiber. If we know that molecules in a particular voxel diffuse principally in one direction we can make the assumption that the majority of the fibers in this area are going parallel to that direction. The recent development of diffusion tensor imaging (DTI) enables diffusion to be measured in multiple directions and the fractional anisotropy in each direction to be calculated for each voxel. This enables researchers to make brain maps of fiber directions to examine the connectivity of different regions in the brain (using tractography) or to examine areas of neural degeneration and demyelination in diseases like Multiple Sclerosis. Another application of diffusion MRI is diffusion-weighted imaging (DWI). Following an ischemic stroke, DWI is highly sensitive to the changes occurring in the lesion. It is speculated that increases in restriction (barriers) to water diffusion, as a result of cytotoxic edema (cellular swelling), is responsible for the increase in signal on a DWI scan. The DWI enhancement appears within 5–10 minutes of the onset of stroke symptoms (as compared with computed tomography, which often does not detect changes of acute infarct for up to 4–6 hours) and remains for up to two weeks. Coupled with imaging of cerebral perfusion, researchers can highlight regions of "perfusion/diffusion mismatch" that may indicate regions capable of salvage by reperfusion therapy. Like many other specialized applications, this technique is usually coupled with a fast image acquisition sequence, such as echo planar imaging sequence.

Magnetization Transfer MRI: Magnetization transfer (MT) refers to the transfer of longitudinal magnetization from free water protons to hydration water protons in NMR and MRI. In magnetic resonance imaging of molecular solutions, such as protein solutions, two types of water molecules, free (bulk) and hydration (bound), are found. Free water protons have faster average rotational frequency and hence less fixed water molecules that may cause local field inhomogeneity. Because of this uniformity, most free water protons have resonance frequency lying narrowly around the normal proton resonance frequency of 63 MHz (at 1.5 teslas). This also results in slower transverse magnetization dephasing and hence longer T_2. Conversely, hydration water molecules are slowed down by interaction with solute molecules and hence create field inhomogeneities that lead to wider resonance frequency spectrum. In free liquids, protons, which may be viewed classically as small magnetic dipoles, exhibit translational and rotational motions.

These moving dipoles disturb the surrounding magnetic field however on long enough time-scales (which may be nanoseconds) the average field caused by the motion of protons is zero. This is known as "motional averaging" or narrowing and is characteristic of protons moving freely in liquid. On the other hand, protons bound to macromolecules, such as proteins, tend to have a fixed orientation and so the average magnetic field in close proximity to such structures does not average to zero. The result is a spatial pattern in the magnetic field that gives rise to a residual dipolar coupling (range of precession frequencies) for the protons experiencing the magnetic field.

The wide frequency distribution appears as a broad spectrum that may be several kHz wide. The net signal from these protons disappears very quickly, in inverse proportion to the width, due to the loss of coherence of the spins, i.e. T2 relaxation. Due to exchange mechanisms, such as spin transfer or proton chemical exchange, the (incoherent) spins bound to the macromolecules continually switch places with (coherent) spins in the bulk media and establish a dynamic equilibrium. Magnetization transfer: Although there is no measurable signal from the bound spins, or the bound spins that exchange into the bulk media, their longitudinal magnetization is preserved and may recover only via the relatively slow process of T1 relaxation. If the longitudinal magnetization of just the bound spins can be altered, then the effect can be measured in the spins of the bulk media due to the exchange processes. The magnetization transfer sequence applies RF saturation at a frequency that is far off resonance for the narrow line of bulk water but still on resonance for the bound protons with a spectral linewidth of kHz. This causes saturation of the bound spins which exchange into the bulk water, resulting in a loss of longitudinal magnetization and hence signal decrease in the bulk water. This provides an indirect measure of macromolecular content in tissue. Implementation of magnetization transfer involves choosing suitable frequency offsets and pulse shapes to saturate the bound spins sufficiently strongly, within the safety limits of specific absorption rate for RF irradiation.

T1ρ (T1rho): Molecules have a kinetic energy that is a function of the temperature and is expressed as translational and rotational motions, and by collisions between molecules. The moving dipoles disturb the magnetic field but are often extremely rapid so that the average effect over a long time-scale may be zero. However, depending on the time-scale, the interactions between the dipoles do not always average away. At the slowest extreme the interaction time is effectively infinite and occurs where there are large, stationary field disturbances (e.g. a metallic implant). In this case the loss of coherence is described as a "static dephasing". T2* is a measure of the loss of coherence in an ensemble of spins that include all interactions (including static dephasing). T2 is a measure of the loss of coherence that excludes static dephasing, using an RF pulse to reverse the slowest types of dipolar interaction. There is in fact a continuum of interaction time-scales in a given biological sample and the properties of the refocusing RF pulse can be tuned to refocus more than just static dephasing. In general, the rate of decay of an ensemble of spins is a function of the interaction times and also the power of the RF pulse. This type of decay, occurring under the influence of RF, is known as T1ρ. It is similar to T2 decay but with some slower dipolar interactions refocused as well as the static interactions.

Fluid Attenuated Inversion Recovery (FLAIR): Fluid Attenuated Inversion Recovery (FLAIR) is an inversion-recovery pulse sequence used to null signal from fluids. For example, it can be used in brain imaging to suppress cerebrospinal fluid (CSF) so as to bring out the periventricular hyperintense lesions, such as multiple sclerosis (MS) plaques. By carefully choosing the inversion time TI (the time between the inversion and excitation pulses), the signal from any particular tissue can be suppressed.

Magnetic Resonance Angiography

Magnetic resonance angiography (MRA) generates pictures of the arteries to evaluate them for stenosis (abnormal narrowing) or aneurysms (vessel wall dilatations, at risk of rupture). MRA is often used to evaluate the arteries of the neck and brain, the thoracic and abdominal aorta, the renal arteries, and the legs (called a "run-off"). A variety of techniques can be used to generate the pictures, such as administration of a paramagnetic contrast agent (gadolinium) or using a technique known as "flow-related enhancement" (e.g. 2D and 3D time-of-flight sequences), where most of the signal on an image is due to blood that recently moved into that plane, see also FLASH MRI. Techniques involving phase accumulation (known as phase contrast angiography) can also be used to generate flow velocity maps easily and accurately. Manetic resonance venography (MRV) is a similar procedure that is used to image veins. In this method, the tissue is now excited inferiorly, while signal is gathered in the plane immediately superior to the excitation plane—thus imaging the venous blood that recently moved from the excited plane.

Magnetic Resonance Gated Intracranial CSF Dynamics (MR-GILD): Magnetic resonance gated intracranial cerebrospinal fluid (CSF) or liquor dynamics (MR-GILD) technique is an MR sequence based on bipolar gradient pulse used to demonstrate CSF pulsatile flow in ventricles, cisterns, aqueduct of Sylvius and entire intracranial CSF pathway. It is a method for analyzing CSF circulatory system dynamics in patients with CSF obstructive lesions such as normal pressure hydrocephalus. It also allows visualization of both arterial and venous pulsatile blood flow in vessels without use of contrast agents.

Magnetic Resonance Spectroscopy: Magnetic resonance spectroscopy (MRS) is used to measure the levels of different metabolites in body tissues. The MR signal produces a spectrum of resonances that correspond to different molecular arrangements of the isotope being "excited". This signature is used to diagnose certain metabolic disorders, especially those affecting the brain, and to provide information on tumor metabolism. Magnetic resonance spectroscopic imaging (MRSI) combines both spectroscopic and imaging methods to produce spatially localized spectra from within the sample or patient. The spatial resolution is much lower (limited by the available SNR), but the spectra in each voxel contains information about many metabolites. Because the available signal is used to encode spatial and spectral information, MRSI requires high SNR achievable only at higher field strengths (3 T and above).

Functional MRI: Functional MRI (fMRI) measures signal changes in the brain that are due to changing neural activity. The brain is scanned at low resolution but at a rapid rate (typically once every 2–3 seconds). Increases in neural activity cause changes in the MR signal via T_2 changes; this mechanism is referred to as the BOLD (blood-oxygen-level dependent) effect. Increased neural activity causes an increased demand for oxygen, and the vascular system actually overcompensates for this, increasing the amount of oxygenated hemoglobin relative to deoxygenated hemoglobin. Because deoxygenated hemoglobin attenuates the MR signal, the vascular response leads to a signal increase that is related to the neural activity. The precise nature of the relationship between neural activity and the BOLD signal is a subject of current research. The BOLD effect also allows for the generation of high resolution 3D maps of the venous vasculature within neural tissue. While BOLD signal is the most common method employed for neuroscience studies in human subjects, the flexible nature of MR imaging provides means to sensitize the signal to other aspects of the blood supply. Alternative techniques employ arterial spin labeling (ASL) or weight the MRI signal by cerebral blood flow (CBF) and cerebral blood volume (CBV). The CBV method requires injection of a class of MRI contrast agents that are now in human clinical trials. Because this method has been shown to be far more sensitive than the BOLD technique in preclinical studies, it may potentially expand the role of fMRI in clinical applications. The CBF method provides more quantitative information than the BOLD signal, albeit at a significant loss of detection sensitivity.

Real-time MRI: Real-time MRI refers to the continuous monitoring ("filming") of moving objects in real time. While many different strategies have been developed over the past two decades, a recent development reported a real-time MRI technique based on radial FLASH that yields a temporal resolution of 20 to 30 milliseconds for images with an in-plane resolution of 1.5 to 2.0 mm. The new method promises to add important information about diseases of the joints and the heart. In many cases MRI examinations may become easier and more comfortable for patients.

Interventional MRI: The lack of harmful effects on the patient and the operator make MRI well-suited for "interventional radiology", where the images produced by a MRI scanner are used to guide minimally invasive procedures. Of course, such procedures must be done without *any* ferromagnetic instruments. A specialized growing subset of interventional MRI is that of intraoperative MRI in which the MRI is used in the surgical process. Some specialized MRI systems have been developed that allow imaging concurrent with the surgical procedure. More typical, however, is that the surgical procedure is temporarily interrupted so that MR images can be acquired to verify the success of the procedure or guide subsequent surgical work.

Radiation Therapy Simulation

Because of MRI's superior imaging of soft tissues, it is now being utilized to specifically locate tumors within the body in preparation for radiation therapy treatments. For therapy simulation, a patient is placed in specific, reproducible, body position and scanned. The MRI system then computes the precise location, shape and orientation of the tumor mass, correcting for any spatial distortion inherent in the system. The patient is then marked or tattooed with points that, when combined with the specific body position, permits precise triangulation for radiation therapy.

Current Density Imaging: Current density imaging (CDI) endeavors to use the phase information from images to reconstruct current densities within a subject. Current density imaging works because electrical currents generate magnetic fields, which in turn affect the phase of the magnetic dipoles during an imaging sequence.

Magnetic Resonance Guided Focused Ultrasound: In MRgFUS therapy, ultrasound beams are focused on a tissue—guided and controlled using MR thermal imaging—and due to the significant energy deposition at the focus, temperature within the tissue rises to more than 65 °C (150 °F), completely destroying it. This technology can achieve precise ablation of diseased tissue. MR imaging provides a three-dimensional view of the target tissue, allowing for precise focusing of ultrasound energy. The MR imaging provides quantitative, real-time, thermal images of the treated area. This allows the physician to ensure that the temperature generated during each cycle of ultrasound energy is sufficient to cause thermal ablation within the desired tissue and if not, to adapt the parameters to ensure effective treatment. This allows the physician to ensure that the temperature generated during each cycle of ultrasoound energy is sufficient to cause thermal ablation within the desired tissue and if not, to adapt the parameters to ensure effective treatment.

Multinuclear Imaging: Hydrogen is the most frequently imaged nucleus in MRI because it is present in biological tissues in great abundance, and because its high gyromagnetic ratio gives a strong signal. However, any nucleus with a net nuclear spin could potentially be imaged with MRI. Such nuclei include helium-3, carbon-13, fluorine-19, oxygen-17, sodium-23, phosphorus-31 and xenon-129. ^{23}Na and ^{31}P are naturally abundant in the body, so can be imaged directly. Gaseous isotopes such as ^{3}He or ^{129}Xe must be hyperpolarized and then inhaled as their nuclear density is too low to yield a useful signal under normal conditions. ^{17}O and ^{19}F can be administered in sufficient quantities in liquid form (e.g. ^{17}O-water) that hyperpolarization is not a necessity. Multinuclear imaging is primarily a research technique at present. However, potential applications include functional imaging and imaging of organs poorly seen on ^{1}H MRI (e.g. lungs and bones) or as alternative contrast agents. Inhaled hyperpolarized ^{3}He can be used to image the distribution of air spaces within the lungs. Injectable solutions containing ^{13}C or stabilized bubbles of hyperpolarized ^{129}Xe have been studied as contrast agents for angiography and perfusion imaging. ^{31}P can potentially provide information on bone density and structure, as well as functional imaging of the brain.

Susceptibility Weighted Imaging (SWI): Susceptibility weighted imaging (SWI), is a new type of contrast in MRI different from spin density, T_1, or T_2 imaging. This method exploits the susceptibility differences between tissues and uses a fully velocity compensated, three dimensional, RF spoiled, high-resolution, 3D gradient echo scan. This special data acquisition and image processing produces an enhanced contrast magnitude image very sensitive to venous blood, hemorrhage and iron storage. It is used to enhance the detection and diagnosis of tumors, vascular and neurovascular diseases (stroke and hemorrhage, multiple sclerosis, Alzheimer's), and also detects traumatic brain injuries that may not be diagnosed using other methods

Other Specialized MRI Techniques: Field of research and new methods and variants are often published when they are able to get better results in specific fields. Examples of these recent improvements are T_2-weighted turbo spin-echo (T_2 TSE MRI), double inversion recovery MRI (DIR-MRI) or phase-sensitive inversion recovery MRI (PSIR-MRI), all of them able to improve imaging of the brain lesions. Another example is MP-RAGE (magnetization-prepared rapid acquisition with gradient echo), which improves images of multiple sclerosis cortical lesions.

Portable Instruments

Portable magnetic resonance instruments are available for use in education and field research. Using the principles of Earth's field NMR, they have no powerful polarizing magnet, so that such instruments can be small and inexpensive. Some can be used for both EFNMR spectroscopy and MRI imaging. The low strength of the Earth's field results in poor signal to noise ratios, requiring long scan times to capture spectroscopic data or build up MRI images. Research with atomic magnetometers have discussed the possibility for cheap and portable MRI instruments without the large magnet.

MRI versus CT

A computed tomography (CT) scanner uses X-rays, a type of ionizing radiation, to acquire its images, making it a good tool for examining tissue composed of elements of a higher atomic number than the tissue surrounding them, such as bone and calcifications (calcium based) within the body (carbon based flesh), or of structures (vessels, bowel). MRI, on the other hand, uses non-ionizing radio frequency (RF) signals to acquire its images and is best suited for non-calcified tissue, though MR images can also be acquired from bones and teeth as well as fossils. CT may be enhanced by use of contrast agents containing elements of a higher atomic number than the surrounding flesh such as iodine or barium. Contrast agents for MRI have paramagnetic properties, e.g., gadolinium and manganese. Both CT and MRI scanners are able to generate multiple two-dimensional cross-sections (slices) of tissue and three-dimensional reconstructions. Unlike CT, which uses only X-ray attenuation to generate image contrast, MRI has a long list of properties that may be used to generate image contrast. By variation of scanning parameters, tissue contrast can be altered and enhanced in various ways to detect different features.

MRI can generate cross-sectional images in any plane (including oblique planes). In the past, CT was limited to acquiring images in the axial (or near axial) plane. The scans used to be called Computed *Axial* Tomography scans (CAT scans). However, the development of multi-detector CT scanners with near-isotropic resolution, allows the CT scanner to produce data that can be retrospectively reconstructed in any plane with minimal loss of image quality. For purposes of tumor detection and identification in the brain, MRI is generally superior. However, in the case of solid tumors of the abdomen and chest, CT is often preferred due to less motion artifact. Furthermore, CT usually is more widely available, faster, less expensive, and may be less likely to require the person to be sedated or anesthetized. MRI is also best suited for cases when a patient is to undergo the exam several times successively in the short term, because, unlike CT, it does not expose the patient to the hazards of ionizing radiation.

Economics of MRI

MRI equipment is expensive. 1.5 tesla scanners often cost between $1 million and $1.5 million USD. 3.0 tesla scanners often cost between $2 million and $2.3 million USD. Construction of MRI suites can cost up to $500,000 USD, or more, depending on project scope. MRI scanners have been significant sources of revenue for healthcare providers in the US. This is because of favourable reimbursement rates from insurers and federal government programmes. Insurance reimbursement is provided in two components, an equipment charge for the actual performance of the MRI scan and professional charge for the radiologist's review of the images and/or data. In the US Northeast, an equipment charge might be $3,500 and a professional charge might be $350 although the actual fees received by the equipment owner and interpreting physician are often significantly less and depend on the rates negotiated with insurance companies or determined by governmental action as in the Medicare Fee Schedule. For example, an orthopedic surgery group in Illinois billed a charge of $1,116 for a knee MRI in 2007 but the Medicare reimbursement in 2007 was only $470.91. Many insurance companies require preapproval of an MRI procedure as a condition for coverage. In the US, the Deficit Reduction Act of 2007 significantly reduced reimbursement rates paid by federal insurance programmes for the equipment component of many scans, shifting the economic landscape. Many private insurers have followed suit.

Safety

A number of features of MRI scanning can give rise to risks.

These include:

- Powerful magnetic fields
- Cryogenic liquids
- Noise
- Claustrophobia

In addition, in cases where MRI contrast agents are used, these also typically have associated risks.

Magnetic Field

Most forms of medical or biostimulation implants are generally considered contraindications for MRI scanning. These include pacemakers, vagus nerve stimulators, implantable cardioverter-defibrillators, loop recorders, insulin pumps, cochlear implants, deep brain stimulators and capsules retained from capsule endoscopy. Patients are therefore always asked for complete information about all implants before entering the room for an MRI scan. Several deaths have been reported in patients with pacemakers who have undergone MRI scanning without appropriate precautions. To reduce such risks, implants are increasingly being developed to make them able to be safely scanned, and specialized protocols have been developed to permit the safe scanning of selected implants and pacing devices. Ferromagnetic foreign bodies such as shell fragments, or metallic implants such as surgical prostheses and aneurysm clips are also potential risks. Interaction of the magnetic and radio frequency fields with such objects can lead to trauma due to movement of the object in the magnetic field or thermal injury from radio-frequency induction heating of the object. Titanium and its alloys are safe from movement from the magnetic field. In the United States a classification system for implants and ancillary clinical devices has been developed by ASTM International and is now the standard supported by the US Food and Drug Administration:

- *MR-Safe* — The device or implant is completely non-magnetic, non-electrically conductive, and non-RF reactive, eliminating all of the primary potential threats during an MRI procedure.
- *MR-Conditional* — A device or implant that may contain magnetic, electrically conductive or RF-reactive components that is safe for operations in proximity to the MRI, provided the conditions for safe operation are defined and observed (such as 'tested safe to 1.5 teslas' or 'safe in magnetic fields below 500 gauss in strength').
- *MR-Unsafe* — Nearly self-explanatory, this category is reserved for objects that are significantly ferromagnetic and pose a clear and direct threat to persons and equipment within the magnet room.

The very high strength of the magnetic field can also cause "missile-effect" accidents, where ferromagnetic objects are attracted to the center of the magnet, and there have been incidences of injury and death. To reduce the risks of projectile accidents, ferromagnetic objects and devices are typically prohibited in proximity to the MRI scanner and patients undergoing MRI examinations are required to remove all metallic objects, often by changing into a gown or scrubs and ferromagnetic detection devices are used by some sites There is no evidence for biological harm from even very powerful static magnetic fields

Radio Frequency Energy

A powerful radio transmitter is needed for excitation of proton spins. This can heat the body to the point of risk of hyperthermia in patients, particularly in obese patients or those with thermoregulation disorders. Several countries have issued restrictions on the maximum specific absorption rate that a scanner may produce.

Peripheral Nerve Stimulation (PNS)

The rapid switching on and off of the magnetic field gradients is capable of causing nerve stimulation. Volunteers report a twitching sensation when exposed to rapidly switched fields, particularly in their extremities. The reason the peripheral nerves are stimulated is that the changing field increases with distance from the center of the gradient coils (which more or less coincides with the center of the magnet). Note however that when imaging the head, the heart is far off-center and induction of even a tiny current into the heart must be avoided at all costs. Although PNS was not a problem for the slow, weak gradients used in the early days of MRI, the strong, rapidly switched gradients used in techniques such as EPI, fMRI, diffusion MRI, etc. are indeed capable of inducing PNS. American and European regulatory agencies insist that manufacturers stay below specified dB/dt limits (dB/dt is the change in field per unit time) or else prove that no PNS is induced for any imaging sequence. As a result of dB/dt limitation, commercial MRI systems cannot use the full rated power of their gradient amplifiers.

Acoustc Noise

Switching of field gradients causes a change in the Lorentz force experienced by the gradient coils, producing minute expansions and contractions of the coil itself. As the switching is typically in the audible frequency range, the resulting vibration produces loud noises (clicking or beeping). This is most marked with high-field machines and rapid-imaging techniques in which sound intensity can reach 120 dB(A) (equivalent to a jet engine at take-off), and therefore appropriate ear protection is essential for anyone inside the MRI scanner room during the examination.

Cryogens

As described in Physics of Magnetic Resonance Imaging, many MRI scanners rely on cryogenic liquids to enable superconducting capabilities of the electromagnetic coils within. Though the cryogenic liquids used are non-toxic, their physical properties present specific hazards. An unintentional shut-down of a superconducting electromagnet, an event known as "quench", involves the rapid boiling of liquid helium from the device. If the rapidly expanding helium cannot be dissipated through an external vent, sometimes referred to as 'quench pipe', it may be released into the scanner room where it may cause displacement of the oxygen and present a risk of asphyxiation. Oxygen deficiency monitors are usually used as a safety precaution.

Liquid helium, the most commonly used cryogen in MRI, undergoes near explosive expansion as it changes from liquid to a gaseous state. The use of an Oxygen monitor is important for two reasons: 1. Keeping the O2 levels safe for patient/physicians and; 2. Making sure the MRI Scanner is working properly. Rooms built in support of superconducting MRI equipment should be equipped with pressure relief mechanisms and an exhaust fan, in addition to the required quench pipe. Since a quench results in rapid loss of all cryogens in the magnet, recommissioning the magnet is expensive and time-consuming. Spontaneous quenches are uncommon, but may also be triggered by equipment malfunction, improper cryogen fill technique, contaminants inside the cryostat, or extreme magnetic or vibrational disturbances.

Contrast Agents

The most commonly used intravenous contrast agents are based on chelates of gadolinium. In general, these agents have proved safer than the iodinated contrast agents used in X-ray radiography or CT. Anaphylactoid reactions are rare, occurring in approx. 0.03–0.1%. Of particular interest is the lower incidence of nephrotoxicity, compared with iodinated agents, when given at usual doses—this has made contrast-enhanced MRI scanning an option for patients with renal impairment, who would otherwise not be able to undergo contrast-enhanced CT. Although gadolinium agents have proved useful for patients with renal impairment, in patients with severe renal failure requiring dialysis there is a risk of a rare but serious illness, nephrogenic systemic fibrosis, that may be linked to the use of certain gadolinium-containing agents. The most frequently linked is gadodiamide, but other agents have been linked too. Although a causal link has not been definitively established, current guidelines in the United States are that dialysis patients should only receive gadolinium agents where essential, and that dialysis should be performed as soon as possible after the scan to remove the agent from the body promptly. In Europe, where more gadolinium-containing agents are available, a classification of agents according to potential risks has been released. Recently a new contrast agent named gadoxetate, brand name Eovist (US) or Primovist (EU), was approved for diagnostic use: this has the theoretical benefit of a dual excretion path.

Pregnancy

No effects of MRI on the fetus have been demonstrated. In particular, MRI avoids the use of ionizing radiation, to which the fetus is particularly sensitive. However, as a precaution, current guidelines recommend that pregnant women undergo MRI only when essential. This is particularly the case during the first trimester of pregnancy, as organogenesis takes place during this period. The concerns in pregnancy are the same as for MRI in general, but the fetus may be more sensitive to the effects—particularly to heating and to noise. However, one additional concern is the use of contrast agents; gadolinium compounds are known to cross the placenta and enter the fetal bloodstream, and it is recommended that their use be avoided. Despite these concerns, MRI is rapidly growing in importance as a way of diagnosing and monitoring congenital defects of the fetus because it can provide more diagnostic information than ultrasound and it lacks the ionizing radiation of CT. MRI without contrast agents is the imaging mode of choice for pre-surgical, in-utero diagnosis and evaluation of fetal tumors, primarily teratomas, facilitating open fetal surgery, other fetal interventions, and planning for procedures (such as the EXIT procedure) to safely deliver and treat babies whose defects would otherwise be fatal.

Claustrophobia and Discomfort

Due to the construction of some MRI scanners, they can be potentially unpleasant to lie in. Older models of closed bore MRI systems feature a fairly long tube or tunnel. The part of the body being imaged must lie at the center of the magnet, which is at the absolute center of the tunnel. Because scan times on these older scanners may be long (occasionally up to 40 minutes for the entire procedure), people with even mild claustrophobia are sometimes unable to tolerate an MRI scan without management. Modern scanners may have larger bores (up to 70 cm) and scan times are shorter. This means that claustrophobia is less of an issue, and many patients now find MRI an innocuous and easily tolerated procedure. Nervous patients may still find the following strategies helpful:

- Advance preparation
  - visiting the scanner to see the room and practice lying on the table
  - visualization techniques
  - chemical sedation
  - general anesthesia
- Coping while inside the scanner
  - holding a "panic button"
  - closing eyes as well as covering them (e.g. washcloth, eye mask)

- o listening to music on headphones or watching a movie, using mirror-glasses and a projection screen or via a Head-mounted display, while in the machine.

Alternative scanner designs, such as open or upright systems, can aso be helpful where these are available. Though open scanners have increased in popularity, they produce inferior scan quality because they operate at lower magnetic fields than closed scanners. However, commercial 1.5 tesla open systems have recently become available, providing much better image quality than previous lower field strength open models. For babies and young children chemical sedation or general anesthesia are the norm, as these subjects cannot be instructed to hold still during the scanning session. Obese patients and pregnant women may find the MRI machine to be a tight fit. Pregnant women may also have difficulty lying on their backs for an hour or more without moving.

Guidance

Safety issues, including the potential for biostimulation device interference, movement of ferromagnetic bodies, and incidental localized heating, have been addressed in the American College of Radiology's *White Paper on MR Safety,* which was originally published in 2002 and expanded in 2004. The *ACR White Paper on MR Safety* has been rewritten and was released early in 2007 under the new title *ACR Guidance Document for Safe MR Practices.* In December 2007, the Medicines in Healthcare product Regulation Agency (MHRA), a UK healthcare regulatory body, issued their *Safety Guidelines for Magnetic Resonance Imaging Equipment in Clinical Use.* In February 2008, the Joint Commission, a US healthcare accrediting organization, issued a Sentinel Event Alert #38, their highest patient safety advisory, on MRI safety issues. In July 2008, the United States Veterans Administration, a federal governmental agency serving the healthcare needs of former military personnel, issued a substantial revision to their *MRI Design Guide,* which includes physical or facility safety considerations.

The European Physical Agents Directive

The European Physical Agents (Electromagnetic Fields) Directive is legislation adopted in European legislature. Originally scheduled to be required by the end of 2008, each individual state within the European Union must include this directive in its own law by the end of 2012. Some member nations passed complying legislation and are now attempting to repeal their state laws in expectation that the final version of the EU Physical Agents Directive will be substantially revised prior to the revised adoption date. The directive applies to occupational xposure to electromagnetic fields (not medical exposure) and was intended to limit workers' acute exposure to strong electromagnetic fields, as may be found near electricity substations, radio or television transmitters or industrial equipment. However, the regulations impact significantly on MRI, with separate sections of the regulations limiting exposure to static magnetic fields, changing magnetic fields and radio frequency energy. Field strength limits are given, which may not be exceeded. An employer may commit a criminal offense by allowing a worker to exceed an exposure limit, if that is how the Directive is implemented in a particular member state.

The Directive is based on the international consensus of established effects of exposure to electromagnetic fields, and in particular the advice of the European Commissions's advisor, the International Commission on Non-Ionizing Radiation Protection (ICNIRP). The aims of the Directive, and the ICNIRP guidelines it is based on, are to prevent exposure to potentially harmful fields. The actual limits in the Directive are very similar to the limits advised by the Institute of Electrical and Electronics Engineers, with the exception of the frequencies produced by the gradient coils, where the IEEE limits are significantly higher.

Many Member States of the EU already have either specific EMF regulations or (as in the UK) a general requirement under workplace health and safety legislation to protect workers against electromagnetic fields. In almost all cases the existing regulations are aligned with the ICNIRP limits so that the Directive should, in theory, have little impact on any employer already meeting their legal responsibilities.

The introduction of the Directive has brought to light an existing potential issue with occupational exposures to MRI fields. There are at present very few data on the number or types of MRI practice that might lead to exposures in excess of the levels of the Directive. There is a justifiable concern amongst MRI practitioners that if the Directive were to be enforced more vigorously than existing legislation, the use of MRI might be restricted, or working practices of MRI personnel might have to change. In the initial draft a limit of static field strength to 2 T was given. This has since been removed from the regulations, and whilst it is unlikely to be restored as it was without a strong justification, some restriction on static fields may be reintroduced after the matter has been considered more fully by ICNIRP. The effect of such a limit might be to restrict the installation, operation and maintenance of MRI scanners with magnets of 2 T and stronger. As the increase in field strength has been instrumental in developing higher resolution and higher performance scanners, this would be a significant step back.

This is why it is unlikely to happen without strong justification. Individual government agencies and the European Commission have now formed a working group to examine the implications on MRI and to try to address the issue of occupational exposures to electromagnetic fields from MRI.

Three-Dimensional (3D) Image Reconstruction

The Principle

Because contemporary MRI scanners offer isotropic, or near isotropic, resolution, display of images does not need to be restricted to the conventional axial images. Instead, it is possible for a software programme to build a volume by 'stacking' the individual slices one on top of the other. The programme may then display the volume in an alternative manner.

3D Rendering Techniques

Surface rendering; A threshold value of greyscale density is chosen by the operator (e.g. a level that corresponds to fat). A threshold level is set, using edge detection image processing algorithms. From this, a 3-dimensional model can be constructed and displayed on screen. Multiple models can be constructed from various different thresholds, allowing different colors to represent each anatomical component such as bone, muscle, and cartilage. However, the interior structure of each element is not visible in this mode of operation.

Volume rendering: Surface rendering is limited in that it only displays surfaces that meet a threshold density, and only displays the surface closest to the imaginary viewer. In volume rendering, transparency and colors are used to allow a better representation of the volume to be shown in a single image - e.g. the bones of the pelvis could be displayed as semi-transparent, so that even at an oblique angle, one part of the image does not conceal another.

Image Segmentation

Where different structures have similar threshold density, it can become impossible to separate them simply by adjusting volume rendering parameters. The solution is called segmentation, a manual or automatic procedure that can remove the unwanted structures from the image.

2003 Nobel Prize

Reflecting the fundamental importance and applicability of MRI in medicine, Paul Lauterbur of the University of Illinois at Urbana-Champaign and Sir Peter Mansfield of the University of Nottingham were awarded the 2003 Nobel Prize in Physiology or Medicine for their "discoveries concerning magnetic resonance imaging". The Nobel citation acknowledged Lauterbur's insight of using magnetic field gradients to determine spatial localization, a discovery that allowed rapid acquisition of 2D images. Mansfield was credited with introducing the mathematical formalism and developing techniques for efficient gradient utilization and fast imaging. The actual research that won the prize was done almost 30 years before, while Paul Lauterbur was at Stony Brook University in New York. The award was vigorously protested by Raymond Vahan Damadian, founder of FONAR Corporation, who claimed that he invented the MRI, and that Lauterbur and Mansfield had merely refined the technology. An ad hoc group, called "The Friends of Raymond Damadian", took out full-page advertisements in the *New York Times* and *The Washington Post* entitled "The Shameful Wrong That Must Be Righted", demanding that he be awarded at least a share of the Nobel Prize. Also, even earlier, in the Soviet Union, Vladislav Ivanov filed (in 1960) a document with the USSR State Committee for Inventions and Discovery at Leningrad for a Magnetic Resonance Imaging device, although this was not approved until the 1970s. In a letter to *Physics Today*, Herman Carr pointed out his own even earlier use of field gradients for one-dimensional MR imaging.

Virtual Particle

In physics, a virtual particle is a particle that exists for a limited time and space, introducing uncertainty in its energy and momentum due to the uncertainty principle. Because energy and momentum in quantum mechanics are time and space derivative operators, then due to Fourier transforms their spans are inversely proportional to time duration and position spans, respectively Virtual particles exhibit some of the phenomena that real particles do, such as obedience to the conservation laws.

If a single particle is detected, then the consequences of its existence are prolonged to such a degree that it cannot be virtual. Virtual particles are viewed as the quanta that describe fields of the basic force interactions, which cannot be described in terms of real particles. Examples of these are static force fields, such as a simple electric or magnetic field, or any field that exists without excitations that result in its carrying informtion from place to place. Virtual photons are also a major component of antenna near field phenomena and induction fields, which have only very short-range effects, that do not radiate through space with the same range-properties as do electromagnetic wave photons. For example, the energy carried from one winding of a transformer to another, in quantum terms, is carried by virtual photons, not real photons. The virtual particle forms of massless particles, such as photons, do have mass (which may be either positive or negative) and are said to be off mass shell. They are allowed to have mass (which consists of "borrowed energy") because they exist for only a temporary time, which in turn gives them a limited "range". This is in accordance with the uncertainty principle which allows existence of such particles of borrowed energy, so long as their energy, multiplied by the time they exist, is a fraction of Planck's constant. The concept of virtual particles is closely related to the idea of quantum fluctuations. Virtual particles can be thought of as coming into existence as quantities, such as the electric field, fluctuate around their expectation values as required by quantum mechanics.

Properties

The concept of virtual particles arises in the perturbation theory of quantum field theory, an approximation scheme in which interactions (essentially forces) between real particles are calculated in terms of exchanges of virtual particles. Any process involving virtual particles admits a schematic representation known as a Feynman diagram which facilitates the understanding of calculations. A virtual particle is one that does not precisely obey the $m^2c^4 = E^2 - p^2c^2$ relationship for a short time. In other words, its kinetic energy may not have the usual relationship to velocity–indeed, it can be negative.

The probability amplitude for it to exist tends to be canceled out by destructive interference over longer distances and times. A virtual particle can be considered a manifestation of quantum tunnelling. The range of forces carried by virtual particles is limited by the uncertainty principle, which regards energy and time as conjugate variables; thus virtual particles of larger mass have more limited range. There is not a definite line differentiating virtual particles from real particles — the equations of physics just describe particles (which includes both equally). The amplitude that a virtual particle exists interferes with the amplitude for its non-existence; whereas for a real particle the cases of existence and non-existence cease to be coherent with each other and do not interfere any more. In the quantum field theory view, "real particles" are viewed as being detectable excitations of underlying quantum fields. As such, virtual particles are also excitations of the underlying fields, but are detectable only as forces but not particles. They are "temporary" in the sense that they appear in calculations, but are not detected as single particles. Thus, in mathematical terms, they never appear as indices to the scattering matrix, which is to say, they never appear as the observable inputs and outputs of the physical process being modelled. In this sense, virtual particles are an artifact of perturbation theory, and do not appear in a non-perturbative treatment. There are two principal ways in which the notion of virtual particles appears in modern physics. They appear as intermediate terms in Feynman diagrams; that is, as terms in a perturbative calculation. They also appear as an infinite set of states to be summed or integrated over in the calculation of a semi-non-perturbative effect. In the latter case, it is sometimes said that virtual particles cause the effect, or that the effect occurs because of the existence of virtual particles.

Manifestations

There are many observable physical phenomena resulting from interactions involving virtual particles. For bosonic particles which exhibit rest mass when they are free and "real," virtual interactions are characterized by the relatively short range of the force interaction produced by particle exchange. Examples of such short-range interactions are the strong and weak forces, and their associated field bosons. For the gravitational and electromagnetic forces, the zero rest-mass of the associated boson particle permits long-range forces to be mediated by virtual particles. However, in the case of photons, power and information transfer by virtual particles is a relatively short-range phenomenon (existing importantly only within a few wavelengths of the field-disturbance which carries information or transferred power), as for example seen in the characteristically short range of inductive and capacitative effects in the near field zone of coils and antennas. Some field interactions which may be seen in terms of virtual particles are:

- The Coulomb force (static electric force) between electric charges. It is caused by the exchange of virtual photons. In symmetric 3-dimensional space this exchange results in the inverse square law for electric force. Since the photon has no mass, the coulomb potential has an infinite range.
- The magnetic field between magnetic dipoles. It is caused by the exchange of virtual photons. In symmetric 3-dimensional

space this exchange results in the inverse square law for magnetic force. Since the photon has no mass, the magnetic potential has an infinite range.

- The so-called near field of radio antennas, where the magnetic and electric effects of the changing current in the antenna wire and the charge effects of the wire's capacitive charge are detectable, but both of which effects decay with increasing distance from the antenna much more quickly than do the influence of conventional electromagnetic waves, for which E is always equal to cB, and which are composed of real photons.
- The strong nuclear force between quarks is the result of interaction of virtual gluons. The residual of this force outside of quark triplets (neutron and proton) holds neutrons and protons together in nuclei, and is due to virtual mesons such as the pi meson and rho meson.
- The weak nuclear force - it is the result of exchange by virtual W and Z bosons.
- The spontaneous emission of a photon during the decay of an excited atom or excited nucleus; such a decay is prohibited by ordinary quantum mechanics and requires the quantization of the electromagnetic field for its explanation.
- The Casimir effect, where the ground state of the quantized electromagnetic field causes attraction between a pair of electrically neutral metal plates.
- The van der Waals force, which is partly due to the Casimir effect between two atoms,
- Vacuum polarization, which involves pair production or the decay of the vacuum, which is the spontaneous production of particle-antiparticle pairs (such as electron-positron).
- Lamb shift of positions of atomic levels.
- Hawking radiation, where the gravitational field is so strong that it causes the spontaneous production of photon pairs (with black body energy distribution) and even of particle pairs.

Most of these have analogous effects in solid-state physics; indeed, one can often gain a better intuitive understanding by examining these cases. In semiconductors, the roles of electrons, positrons and photons in field theory are replaced by electrons in the conduction band, holes in the valence band, and phonons or vibrations of the crystal lattice. A virtual particle is in a virtual state where the probability amplitude is not conserved. Examples of macroscopic virtual phonons, photons, and electrons in the case of the tunneling process were presented by Günter Nimtz in.

History

Paul Dirac was the first to propose that empty space (a vacuum) can be visualized as consisting of a sea of virtual electron-positron pairs, known as the Dirac sea. The Dirac sea has a direct analog to the electronic band structure in crystalline solids as described in solid state physics. Here, particles correspond to conduction electrons, and antiparticles to holes. A variety of interesting phenomena can be attributed to this structure.

Virtual Particles in Feynman Diagrams

The calculation of scattering amplitudes in theoretical particle physics requires the use of some rather large and complicated integrals over a large number of variables. These integrals do, however, have a regular structure, and may be represented as Feynman diagrams. The appeal of the Feynman diagrams is strong, as it allows for a simple visual presentation of what would otherwise be a rather arcane and abstract formula. In particular, part of the appeal is that the outgoing legs of a Feynman diagram can be associated with real, on-shell particles. Thus, it is natural to associate the other lines in the diagram with particles as well, called the "virtual particles". Mathematically, they correspond to the propagators appearing in the diagram. In the image to the right, the solid lines correspond to real particles (of momentum p_1 and so on), while the dotted line corresponds to a virtual particle carrying momentum k. For example, if the solid lines were to correspond to electrons interacting by means of the electromagnetic interaction, the dotted line would correspond to the exchange of a virtual photon. In the case of interacting nucleons, the dotted line would be a virtual pion. In the case of quarks interacting by means of the strong force, the dotted line would be a virtual gluon, and so on.

It is sometimes said that all photons are virtual photons. This is because the world-lines of photons always resemble the dotted line in the above Feynman diagram: the photon was emitted somewhere (say, a distant star), and then is absorbed somewhere else (say a photoreceptor cell in the eyeball). Furthermore, in a vacuum, a photon experiences no passage of (proper) time between emission and absorption. This statement illustrates the difficulty of trying to distinguish between "real" and "virtual" particles as mathematically they are the same objects and it is only our definition of "reality" which is weak here. In practice, a clear distinction can be made: real photons are detected as individual particles in particle detectors, whereas virtual photons are not directly detected; only their average or side-effects may be noticed, in the form of *forces* or (in modern language) *interactions* between particles. Virtual particles need not be mesons or vector bosons, as in the example above; they may also be fermions. However, in order to preserve quantum numbers, most simple diagrams involving fermion exchange are prohibited. The image to the right shows an allowed diagram, a one-loop diagram. The solid lines correspond to a fermion propagator, the wavy lines to bosons.

Pair Production

In order to conserve the total fermion number of the universe, a fermion cannot be created without also creating its antiparticle; thus many physical processes lead to pair creation. The need for the normal ordering of particle fields in the vacuum can be interpreted by the idea that a pair of virtual particles may briefly "pop into existence", and then annihilate each other a short while later. Thus, virtual particles are often popularly described as coming in pairs, a particle and antiparticle, which can be of any kind. These pairs exist for an extremely short time, and mutually annihilate in short order. In some cases, however, it is possible to boost the pair apart using external energy so that they avoid annihilation and become real particles. This may occur in one of two ways. In an accelerating frame of reference, the virtual particles may appear to be real to the accelerating observer; this is known as the Unruh effect. In short, the vacuum of a stationary frame appears, to the accelerated observer, to be a warm gas of real particles in thermodynamic equilibrium. The Unruh effect is a toy model for understanding Hawking radiation, the process by which black holes evaporate. Another example is pair production in very strong electric fields, sometimes called vacuum decay. If, for example, a pair of atomic nuclei are merged together to very briefly form a nucleus with a charge greater than about 140, (that is, larger than about the inverse of the fine structure constant), the strength of the electric field will be such that it will be energetically favourable to create positron-electron pairs out of the vacuum or Dirac sea, with the electron attracted to the nucleus to annihilate the positive charge.

This pair-creation amplitude was first calculated by Julian Schwinger in 1951. The restriction to particle-antiparticle pairs is actually only necessary if the particles in question carry a conserved quantity, such as electric charge, which is not present in the initial or final state. Otherwise, other situations can arise. For instance, the beta decay of a neutron can happen through the emission of a single virtual, negatively charged W particle that almost immediately decays into a real electron and antineutrino; the neutron turns into a proton when it emits the W particle. The evaporation of a black hole is a process dominated by photons, which are their own antiparticles and are uncharged.

Spontaneous Fission

Spontaneous fission (SF) is a form of radioactive decay characteristic of very heavy isotopes. It is theoretically possible for any atomic nucleus whose mass is greater than or equal to 100 atomic mass units (u), i.e. elements near ruthenium. In practice, however, spontaneous fission is only energetically feasible for atomic masses above 230 u (elements near thorium). The elements most susceptible to spontaneous fission are the high-atomic-number actinide elements, such as mendelevium and lawrencium, and the trans-actinide elements, such as rutherfordium. For uranium and thorium, the spontaneous fission mode of decay does occur, but it is not seen for the majority of radioactive breakdowns, and it is usually neglected except for the exact considerations of branching ratios when determining the activity of a sample containing these elements. As the name suggests, spontaneous fission gives the same result as induced nuclear fission. However, like other forms of radioactive decay, it occurs due to quantum tunneling, without the atom having been struck by a neutron or other particle as in induced nuclear fission. Spontaneous fissions release neutrons as all fissions do, so if a critical mass is present, a spontaneous fission can initiate a chain reaction. Also, radioisotopes for which spontaneous fission is a nonnegligible decay mode may be used as neutron sources; californium-252 (half-life 2.645 years, SF branch ratio 3.09%) is often used for this purpose.

The neutrons may then be used to inspect airline luggage for hidden explosives, to gauge the moisture content of soil in the road construction and building industries, to measure the moisture of materials stored in silos, and in other applications. As long as the fissions give a negligible reduction of the amount of nuclei that can spontaneously fission, this is a Poisson process: for very short time intervals the probability of a spontaneous fission is proportional to the length of time. The spontaneous fission of uranium-238 leaves trails of damage in uranium-bearing minerals as the fission fragments recoil through the crystal structure. These trails, or *fission tracks*, provide the basis for the radiometric dating technique known as fission track dating.

History

The first nuclear fission process discovered was the fission induced by neutrons. Because cosmic rays produce some neutrons, it was difficult to distinguish between induced and spontaneous fission events. Cosmic rays can be reliably shielded by a thick layer of rock or water. The spontaneous fission was identified in 1940 by Soviet physicists Georgy Flyorov and Konstantin Petrzhak by their observations of uranium in the Moscow Metro Dinamo station, 60 metres (200 ft) deep underground.

Neutron Emission

Neutron emission is a type of radioactive decay of atoms containing excess neutrons, in which a neutron is simply ejected from the nucleus. Two examples of isotopes which emit neutrons are helium-5 and beryllium-13. However, the decay of helium-5 is also (by definition) a case of alpha-decay. Many heavy isotopes, most notably californium-252, emit neutrons among the products of a different radioactive decay process, spontaneous fission. Neutrons are absorbed and emitted in the process of nuclear fission, a nuclear chain reaction propagated by neutrons. Delayed neutrons emitted by neutron-rich fission products aid control of nuclear reactors by making reactivity change much slower than it would be for prompt neutrons alone.

Positron Emission

Positron emission or beta plus decay (β^+ decay) is a type of beta decay in which a proton is converted, via the weak force, to a neutron, releasing a positron (the antimatter counterpart of an electron) and a neutrino. Isotopes which undergo this decay and thereby emit positrons include carbon-11, potassium-40, nitrogen-13, oxygen-15, fluorine-18, and iodine-121. Inside protons and neutrons, there are fundamental particles called quarks. The two most common types of quarks are up quarks, which have a charge of $+^2/_3$ and down quarks, with a $-^1/_3$ charge. Quarks arrange themselves in sets of three such that they make protons and neutrons. In a proton, whose charge is +1, there are two up quarks and one down quark. Neutrons, with no charge, have one up quark and two down quarks. Quarks are able to change from up quarks to down quarks.

It is this that causes beta radiation. Positron emission happens when an up quark changes into a down quark. These isotopes are used in positron emission tomography, a technique used for medical imaging. Note that the energy emitted depends on the isotope that is decaying;

the figure of 0.96 MeV applies only to the decay of carbon-11. Isotopes which increase in mass under the conversion of a proton to a neutron, or which decrease by less than $2m_e$, do not spontaneously decay by positron emission. Nuclei which decay by positron emission may also decay by electron capture.

For low-energy decays, electron capture is energetically favoured by $2m_ec^2 = 1.022$ MeV, since the final state has an electron removed rather than a positron added. As the energy of the decay goes up, so does the branching ratio towards positron emission. However, if the energy difference is less than $2m_ec^2$, then positron emission cannot occur and electron capture is the sole decay mode. Certain isotopes (for instance, ^7Be) are stable in galactic cosmic rays, because the electrons are stripped away and the decay energy is too small for positron emission.

Proton Emission

Proton emission (also known as proton radioactivity) is a type of radioactive decay in which a proton is ejected from a nucleus. Proton emission can occur from high-lying excited states in a nucleus following a beta decay, in which case the process is known as beta-delayed proton emission, or can occur from the ground state (or a low-lying isomer) of very proton-rich nuclei, in which case the process is very similar to alpha decay. For a proton to escape a nucleus, the proton separation energy must be negative - the proton is therefore unbound, and tunnels out of the nucleus in a finite time.

Proton emission is not seen in naturally-occurring isotopes; proton emitters can be produced via nuclear reactions, usually utilising some kind of particle accelerator. Although prompt (i.e. not beta-delayed) proton emission was observed from an isomer in Cobalt-53 as early as 1969, no other proton-emitting states were found until 1981, when the proton radioactive ground states of Lutetium-151 and Thulium-147 were observed at experiments at the GSI in West Germany. Research in the field flourished after this breakthrough, and to date more than 25 isotopes have been found to exhibit proton emission.

The study of proton emission has aided the understanding of nuclear deformation, masses and structure, and it is a wonderfully pure example of quantum tunneling. In 2002, the simultaneous emission of two protons was observed from the nucleus Iron-45 in experiments at GSI and GANIL (Grand Accélérateur National d'Ions Lourds, near Caen). In 2005 it was experimentally determined (at the same facility) that Zinc-54 can also undergo double proton decay.

Electron Capture

Electron capture is a process in which a proton-rich nuclide absorbs an inner atomic electron (changing a nuclear proton to a neutron) and simultaneously emits a neutrino. Various photon emissions follow, in order to allow the energy of the atom to fall to the ground state of the new nuclide. Electron capture is the primary decay mode for isotopes that have a relative superabundance of protons in the nucleus, but there is insufficient energy difference between the isotope and its prospective daughter with one less positive charge, for the nuclide to decay by simply emitting a positron.

Electron capture also exists as a viable decay mode for radioactive isotopes that do have enough energy to decay by positron emission, and, in that case, it competes with positron emission. It is sometimes called inverse beta decay, though this term can also refer to the capture of a neutrino through a similar process. If the energy difference between the parent atom and the daughter atom is less than 1.022 MeV, positron emission is forbidden because not enough decay energy is available to allow it, and thus electron capture is the sole decay mode. For example, rubidium-83 (37 protons, 46 neutrons) will decay to krypton-83 (36 protons, 47 neutrons) solely by electron capture (the energy difference, or decay energy, is about 0.9 MeV).

Since the proton is changed to a neutron in electron capture, the number of neutrons increases by 1, the number of protons decreases by 1, and the atomic mass number remains unchanged. By changing the number of protons, electron capture transforms the nuclide into a new element. The atom, although still neutral in charge, now exists in an energetically excited state with the inner shell missing an electron. While transiting to the ground state, the atom will emit an X-ray photon (a type of electromagnetic radiation) and/or Auger electrons, or both. Often the nucleus exists in an excited state as well, and emits a gamma ray in order to reach the ground state energy of the new nuclide just formed.

History

The theory of electron capture was first discussed by Gian-Carlo Wick in a 1934 paper, and then developed by Hideki Yukawa and others. K-electron capture was first observed by Luis Alvarez, in vanadium-48. He reported it in a 1937 paper in the *Physical Review*. Alvarez went on to study electron capture in gallium-67 and other nuclides. Note that it is one of the initial atom's own electrons that is captured, not a new, incoming electron, as might be suggested by the way the above reactions are written. Radioactive isotopes that decay by pure electron capture can, in theory, be inhibited from radioactive decay if they are fully ionized ("stripped" is sometimes used to describe such ions). It is hypothesized that such elements, if formed by the r-process in exploding supernovae, are ejected fully ionized and so do not undergo radioactive decay as long as they do not encounter electrons in outer space. Anomalies in elemental distributions are thought to be partly a result of this effect on electron capture. Chemical bonds can also affect the rate of electron capture to a small degree (in general, less than 1%) depending on the proximity of electrons to the nucleus. For example in ^{7}Be, a difference of 0.9% has been observed between half-lives in metallic and insulating environments. This relatively large effect is due to the fact that beryllium is a small atom whose valence electrons are close to the nucleus. Around the elements in the middle of the periodic table, isotopes that are lighter than stable isotopes of the same element tend to decay through electron capture, while isotopes heavier than the stable ones decay by electron emission.

Neutron Capture

Neutron capture is a kind of nuclear reaction in which an atomic nucleus collides with one or more neutrons and they merge to form a heavier nucleus. Since neutrons have no electric charge they can enter a nucleus more easily than positively charged particles, which are repelled electrostatically. Neutron capture plays an important role in the cosmic nucleosynthesis of heavy elements. In stars it can proceed in two ways - as a rapid process (an r-process) or a slow process (an s-process). Nuclei of masses greater than 56 cannot be formed by thermonuclear reactions (i.e. by nuclear fusion), but can be formed by neutron capture.

Neutron Capture at Small Neutron Flux

At small neutron flux, as in a nuclear reactor, a single neutron is captured by a nucleus. For example when natural gold (^{197}Au) is irradiated by neutrons the isotope ^{198}Au is formed in a highly excited state which then quickly decays to the ground state of ^{198}Au by the emission of γ rays. In this process the mass number (the number of nucleons - both protons and neutrons) increases by one. In terms of a formula this is written ^{197}Au(n,γ)^{198}Au. If thermal neutrons are used we call it thermal capture. The isotope ^{198}Au is a beta emitter that decays into the mercury isotope ^{198}Hg. In this process the atomic number (the number of protons in the nucleus) rises by one. The s-process mentioned above happens in the same way, but inside of stars.

Neutron Capture at High Neutron Flux

The r-process happens inside stars if the neutron flux density is so high that the atomic nucleus has no time to decay via beta emission in between neutron captures. The mass number therefore rises by a large amount while the atomic number (i.e., the element) stays the same. Only afterwards, the highly unstable nuclei decay via many β^- decays to stable or unstable nuclei of high atomic number.

Capture Cross Section

The absorption neutron cross-section of an isotope of a chemical element is the effective cross sectional area that an atom of that isotope presents to absorption, and is a measure of the probability of neutron capture. It is usually measured in barns (b). Absorption cross section is often highly dependent on neutron energy. Two of the most commonly specified measures are the cross-section for thermal neutron absorption, and resonance integral which considers the contribution of absorption peaks at certain neutron energies specific to a particular nuclide, usually above the thermal range, but encountered as neutron moderation slows the neutron down from an original high energy. The thermal energy of the nucleus also has an effect; as temperatures rise, Doppler broadening increases the chance of catching a resonance peak. In particular, the increase in U-238's ability to absorb neutrons at higher temperatures (and to do so without fissioning) is a negative feedback mechanism that helps keep nuclear reactors under control.

Uses

Neutron activation analysis can be used to remotely detect the chemical composition of materials. This is because different elements release different characteristic radiation when they absorb neutrons. This makes t useful in many fields related to mineral exploration and security.

Neutron Absorbers

The most prolific neutron absorbers are the radioactive isotopes of elements that happen to become stable by absorbing one neutron. An example of these is xenon-135 (half life about 9.1 hours), which absorbs a neutron to become the stable isotope xenon-136. Xenon-135 is formed in nuclear reactors when the splitting of uranium-235, uranium-233, or plutonium-239, in a nuclear chain reaction commonly leads to the production of some iodine-135. Iodine-135 soon undergoes nuclear decay, by emitting a beta particle - with a quite short half-life - to produce xenon-135. Other isotopes that are major neutron-absorbers include helium-3, which becomes tritium, a heavy isotope of hydrogen; and boron-10, which splits to become lithium-7 and a helium-4 nucleus. Samarium-149, which is also formed during nuclear chain reactions, is also a highly effective neutron absorber. It also has a very long half-life, and it lasts effectively permanently in the nuclear fuel until it absorbs a neutron, which transmutes it into samarium-150, a stable isotope. Other important neutron absorbers that are used in nuclear reactors for their control rods include the metals cadmium, hafnium, and sometimes the rare earth element, gadolinium, all of which usually consist of mixtures of various isotopes—some of which are excellent neutron-absorbers. Hafnium, one of the last stable elements to be discovered, presents an interesting case. Even though hafnium is a heaver element, its electron configuration makes it practically identical with the element zirconium, and they are always found in the same ores, and in the gem, the zircon. However, their nuclear properties are different in a profound way. Hafnium absorbs neutrons avidly, and it can be used in reactor control rods. However, natural zirconium is practically transparent to neutrons. Thus, zirconium is very desirable for use in the construction of reactors, including in such parts as the metal cladding of the fuel rods which contain either uranium, plutonium, or an alloy of the two metals. Hence, it is quite important to be able to separate the zirconium from the hafnium in their naturally-occurring alloy. This can only be done inexpensively by using modern chemical ion-exchange resins.

Bibliography

Antsiferov, V. V. and G. I. Smirnov: *Coherent Radiation Processes in Plasmas*, Cambridge Univ. Press, Cambridge, 1999.

Bard, A.J.; Faulkner, L.R.: *Electrochemical Methods: Fundamentals and Applications*, John Wiley & Sons, New York, 2005.

Braams, C. and P. Stott: *Nuclear Fusion: Half a Century of Magnetic Confinement Fusion*, IOP Press, NJ, 2002.

Brauman J. I. and Coulson C. A.: *Supplementary Tables of Molecular Orbital Calculations*, Pergamon Press, Oxford, 1965.

Charles F. Kennel, *Convection and Substorms*, Oxford Univ. Press, Oxford, 1995.

Cox, Brian; Forshaw, Jeff: *The Quantum Universe: Everything That Can Happen Does Happen*, Allen Lane, UK, 2011.

Dahl, P. F.: *Superconductivity: Its Historical Roots and Development from Mercury to the Ceramic Oxides*, American Institute of Physics, 1992.

David Chandler: *Introduction to Modern Statistical Mechanics*, Oxford University Press, Oxford, 1987.

Davidson, Ronald: *An Introduction to the Physics of Nonneutral Plasma*, Addison-Wesley, London, 1990.

Ehrenfest, P. and T.: *The Conceptual Foundations of the Statistical Approach in Mechanics*, Cornell University Press, Ithaca, NY, 1959.

Freidberg, J. P.: *Ideal Magnetohydrodynamics*, Plenum Pr., NY, 1987.

Freyd, P.: *Functor Theory*, Princeton University, Princeton, New Jersey, 1960.

Gabriel, P. and M.Zisman:*Category of Fractions and Homotopy Theory*, Springer, Berlin, 1967.

Ghilardi, S. & Zawadowski, M.: *Model Completions: A Categorical Approach to Nonclassical Porpositional Logics*, Dordrecht, Kluwer, 2002.

Gunther Rudiger and Rainer Hollerbach: *The Magnetic Universe: Geophysical and Astrophysical Dynamo Theory* , Wiley-VCH, 2004.

Halmos, P.: *Introduction to Hilbert Space*, AMS Chelsea Publishing, Providence, 1957.

Hatcher, W. S.: *The Logical Foundations of Mathematics*, Pergamon Press, Oxford, 1982.

Hill, Terrell L.: *Introduction to Statistical Thermodynamics*, Dover Publications, New York, 1986.

Ibach H. and H. Lÿÿth: *Solid-State Physics: an Introduction to Principles of Materials Science,* Springer, Berlin; New York, 1996.

Johnstone, P. T.: *Topos Theory*, Academic Press, New York, 1977.

Kallenrode, M.B.: *Space Physics: Plasmas and particles in the Heliosphere and Magnetosheres*, Springer, London, 1998.

Kresin, V. Z. and S. A. Wolf: *Fundamentals of Superconductivity,* Plenum Press, New York, 1990.

Lambek, J. and P. J. Scott: *Introduction to Higher Order Categorical Logic*, Cambridge University Press, Cambridge, 1986.

Leonard F. Burlaga: *Interplanetary Magnetohydrodynamics*, Oxford Univ. Press, Oxford, 1995.

Mehra, J. and H. Rechenberg: *The Historical Development of Quantum Theory*, Springer-Verlag, New York 1982.

Moiseev, S. S.; V. Oraevsky, and V. Pungin: *Non-Linear Instabilities in Plasmas and Hydrodynamics*, IOP Press, NJ, 1999.

Neil Gershenfeld: *The Physics of Information Technology,* Cambridge Univ. Press, Cambridge, 2000.

Plymen, R.J. and P. L. Robinson: *Spinors in Hilbert Space*, Cambridge Univ. Press, Cambridge, 1994.

Popescu, N.: *Abelian Categories with Applications to Rings and Modules,* Academic Press., New York and London, 1973.

Rae, A.: *Quantum Physics: Illusion or Reality?*, Cambridge University Press, Cambridge, 1986.

Redhead, M.L.G.: *Incompleteness, Nonlocality and Realism*, Clarendon Press, Oxford, 1989.

Scott, P. J.: *Some Aspects of Categories in Computer Science*, Amsterdam, North Holland, 2000.

Sklar, L.: *Physics and Chance: Philosophical Issues in the Foundations of Statistical Mechanics*, Cambridge University Press, Cambridge, 1993.

Unruh, W.G.: *Physics Meets Philosophy at the Planck Scale*, Cambridge University Press, Cambridge, 2001.

Vidali, G.: *Superconductivity: The Next Revolution?,* Cambridge University Press, New York, NY, USA, 1993.

Whitaker, A.: *Einstein, Bohr and the Quantum Dilemma*, Cambridge University Press, Cambridge, 1996.

Wilhelmsson, H. and E. Lazzaro: *Reaction-Diffusion in the Physics of Hot Plasmas,* IOP, NJ, 2000.

Index

A

Actinide Fuel, 193, 216.
Alpha Decay, 133, 141, 142, 160, 177, 178, 226, 261, 263, 282, 308.
Aqueous Solutions, 194.
Atomic Spectra, 27, 31.

B

Baryon Asymmetry, 235.
Beta Decay, 131, 134, 135, 140, 141, 144, 145, 146, 154, 155, 156, 157, 159, 247, 250, 252, 267, 302, 304, 305.
Big Bang, 137, 138, 139, 221, 222, 223, 224, 225, 227, 228, 229, 230, 231, 232, 233, 234, 236, 237, 238, 239.
Bioenergy, 175, 176.

C

Chain Reactions, 218, 219, 264, 308.
Cluster Decay, 134, 141, 160, 161, 162, 246.
Contrast Agents, 283, 284, 286, 287, 288, 291, 292.
Cryogens, 290, 291.

D

Dark Energy, 226, 233, 236, 238.
Dissipation, 164, 165.
Double Beta Decay, 144, 154, 156.

E

Electromagnetism, 110, 215.
Electron Capture, 130, 134, 135, 140, 144, 146, 154, 156, 157, 159, 304, 305, 306.
Erotation, 47, 56.
Etranslation, 47, 55, 56.

F

Fission Bombs, 250, 256.
Fossil Fuel, 177, 178, 179, 181, 182, 183, 186, 270.
Fusion Fuels, 201.

G

Gamma Ray, 130, 131, 133, 134, 144, 145, 147, 148, 149, 150, 151, 152, 153, 158, 246, 249, 250, 252, 305.
Global Dimming, 183, 184, 186, 187, 188, 189.

H

Halo Nuclei, 267.
Horizon Problem, 233.

I

Image Sensors, 173.
Infrared Spectroscopy, 55.
Internal Conversion, 130, 134, 140, 151, 157, 158, 159, 160.
Isomeric Transition, 134, 147, 160.

K

Klapdor-Kleingrothaus, 156.
Kurie Plot, 145.

L

Light Interaction, 150.
Liquid Drop Models, 268.
Liquid Fuels, 194.

M

Macrophysics, 82, 83, 88, 95.
Mass Defect, 204, 205, 213.
Mass Deficit, 203, 204.
Metal Fuel, 192, 196.
Microphysics, 82, 83, 88.

N

Neutron Absorbers, 262, 307, 308.
Neutron Capture, 193, 196, 210, 220, 244, 247, 306, 307.
Neutron Emission, 134, 303.
Nuclear Force, 131, 141, 202, 204, 211, 212, 213, 215, 217, 247, 252, 266, 269, 299.
Nuclear Fuel, 176, 190, 191, 192, 194, 195, 196, 198, 239, 243, 244, 245, 246, 247, 249, 253, 254, 270, 308.

O

Oxide Fuel, 191, 192, 193, 198.

P

Photoelectric Effect, 149, 150, 151, 153, 159, 160, 165, 167, 168, 169, 171, 172, 173, 174, 175, 272.
Photovoltaic Effect, 165, 166, 168.
Plum Pudding, 33, 38.
Positron Emission, 134, 144, 156, 303, 304, 305.
Proton Emission, 134, 246, 304.

Q

Q-value, 161, 208.
QUADRISO Fuel, 197.
Quantized Atom, 46.

R

Radio Frequency, 48, 277, 278, 279, 287, 289, 294.
Radioactive Decay, 130, 134, 135, 137, 139, 140, 141, 142, 273, 302, 303, 304, 306.
Reactor Poisons, 243, 244.
Rendering Techniques, 295.

S

Spacecraft, 174, 231.
Speculative Atomism, 25, 44.
Spent Nuclear Fuel, 192, 198, 243.
Spontaneous Fission, 134, 141, 161, 162, 239, 240, 302, 303.
Strong Interaction, 115, 121, 122, 123, 128, 129, 211, 213, 215, 216, 266.

T

Terrestrial Sources, 271, 723.
Three-Step Model, 167, 168.
Toxicity, 142, 291.
TRISO Fuel, 197.

U

Uranium Carbide, 194.
Uranium Nitride, 193.
Uranyl Salts, 194.

V

Virtual Particle, 155, 296, 297, 298, 299, 300, 301.

W

Weak Force, 111, 144, 215, 216, 298, 303.

❑❑❑